Costly Tradeoffs: Reconciling Trade and the Environment

HILARY F. FRENCH

Megan Ryan, *Research Assistant*

Ed Ayres, *Editor*

WORLDWATCH PAPER 113
March, 1993

THE WORLDWATCH INSTITUTE is an independent, nonprofit environmental research organization based in Washington, D.C. Its mission is to foster a sustainable society—in which human needs are met in ways that do not threaten the health of the natural environment or future generations. To this end, the Institute conducts interdisciplinary research on emerging global issues, the results of which are published and disseminated to decisionmakers and the media.

FINANCIAL SUPPORT is provided by the Geraldine R. Dodge Foundation, George Gund Foundation, W. Alton Jones Foundation, John D. and Catherine T. MacArthur Foundation, Andrew W. Mellon Foundation, Curtis and Edith Munson Foundation, Edward John Noble Foundation, Pew Charitable Trusts, Lynn R. and Karl E. Prickett Fund, Public Welfare Foundation, Rockefeller Brothers Fund, Surdna Foundation, Turner Foundation, and Frank Weeden Foundation. The Institute welcomes private donations and bequests.

PUBLICATIONS of the Institute include the annual *State of the World,* which is now published in 27 languages; *Vital Signs,* an annual compendium of the global trends—environmental, economic, and social—that are shaping our future; the *Environmental Alert* book series; and *World Watch* magazine, as well as the *Worldwatch Papers*. For more information on Worldwatch publications, write: Worldwatch Institute, 1776 Massachusetts Ave., N.W., Washington, DC 20036; or FAX (202) 296-7635.

THE WORLDWATCH PAPERS provide in-depth, quantitative and qualitative analysis of the major issues affecting prospects for a sustainable society. The Papers are authored by members of the Worldwatch Institute research staff and reviewed by experts in the field. Published in five languages, they have been used as a concise and authoritative reference by governments, nongovernmental organizations and educational institutions worldwide. For a partial list of available Papers, see page 75.

Library of Congress Catalog Number 93-060169
ISBN 1-878071-14-9

Printed on 100-percent non-chlorine bleached, partially recycled paper.

Table of Contents

Tables and Figures

ACKNOWLEDGEMENTS: I would like to thank John Cavanagh, Steve Charnovitz, Sanford Gaines, Robert Housman, Stewart Hudson, William Snape, and my colleagues at Worldwatch Institute for their comments on preliminary drafts of this paper. Project oversight was provided by Sandra Postel, vice president for research at Worldwatch.

HILARY F. FRENCH is a Senior Researcher at the Worldwatch Institute, where she studies international environmental institutions and politics. She is the author of Worldwatch Paper 107, *After the Earth Summit: the Future of Environmental Governance* and Paper 99, *Green Revolutions: Environmental Reconstruction in Eastern Europe and the Soviet Union,* among other publications. She is also coauthor of four of the Institute's annual *State of the World* reports. She is a graduate of Dartmouth College.

Introduction

Of all the challenges that lie ahead in the race to achieve a sustainable global economy, perhaps none is more riddled with pitfalls than that of international trade. With such looming problems as global warming, deforestation, or biodiversity loss, at least the issues are clearly defined. Trade, on the other hand, cuts across *all* those problems: it plays a central role in deforesting Malaysia and accelerating extinctions of plants and animals in Costa Rica. It exacerbates climate change by increasing the energy requirements of goods transported over longer distances.

But what's most perplexing about global trade is that its implications for the environment confront decision-makers with vexing contradictions and tradeoffs at every turn. The growing mobility of capital through such inventions as the "maquiladoras" free trade zone on the U.S.-Mexican border, for example, undermines environmental standards by allowing polluting industries to escape enforcement rather than clean up their act; yet the same mobility of investment may arguably give poor countries greater revenue to invest in cleaner industries themselves. Freer trade might cause rain forest to be cleared for cash crops; yet by reducing tariffs on processed goods sold abroad, it might also enable developing countries to move beyond heavy dependence on extractive industries such as tropical forest logging and mining.

As a result of these conundrums, the debate over trade and the environment has become a political minefield that reaches into virtually every country, industry, and ecosystem. As much as any other issue, it is this that will test the ability of governments to meet the demands of a world in which "instability" is no longer just a political condition but a biological one.

The dilemmas posed by expanding trade are not only formidable, but urgent—partly because it is expanding so fast. Merchandise trade—including both primary commodities and manufactured products—has grown elevenfold since 1950 and is now tallied at $3.5 trillion (See Figure 1 and Table 1), outpacing the expansion of global output as a whole. Trade in services is also on the rise, as is direct investment in foreign countries.[1]

This rapid expansion poses some obvious threats, such as dumping of toxic waste, destruction of tropical forest, the creation of pollution havens, and—as free trade agreements are concluded—attacks on environmental laws as "non-tariff barriers to trade." Yet it also offers welcome opportunities—such as those discovered by villagers in the Dominican Republic who use photovoltaic cells shipped from the United States to light their homes, or by U.S. auto consumers in the 1970s, whose fuel efficiency was greatly improved by the spur of Japanese automaking competition.[2]

The urgency also arises from the fact that environmental issues have become important factors in regional and global trade negotiations, raising critical issues that must be resolved one way or another within the next few years. At the global level, governments are working to expand upon the 1947 General Agreement on Tariffs and Trade (GATT) that lays the basic ground rules of world trade. Objectives of the ongoing negotiations—collectively known as the "Uruguay Round" after the location where they were launched in 1986—include removing subsidies to agriculture, reducing "non-tariff barriers to trade" such as uneven product standards, and minimizing restrictions on foreign investment. Though free trade enthusiasts view the Uruguay Round as a chance to give the global economy a much-needed jump start, the latest draft of the agreement contains a number of environmentally troubling provisions.[3]

For the twelve members of the European Community (EC), the beginning of 1993 brought the opening of the long-awaited single market—meaning that labor, goods, and capital can now move freely throughout the area. Doing business between Bonn and Brussels is, in theory at least, now as easy as doing business between San Francisco and Seattle—a development

FIGURE 1

World Exports, 1950–91

Trillion Dollars

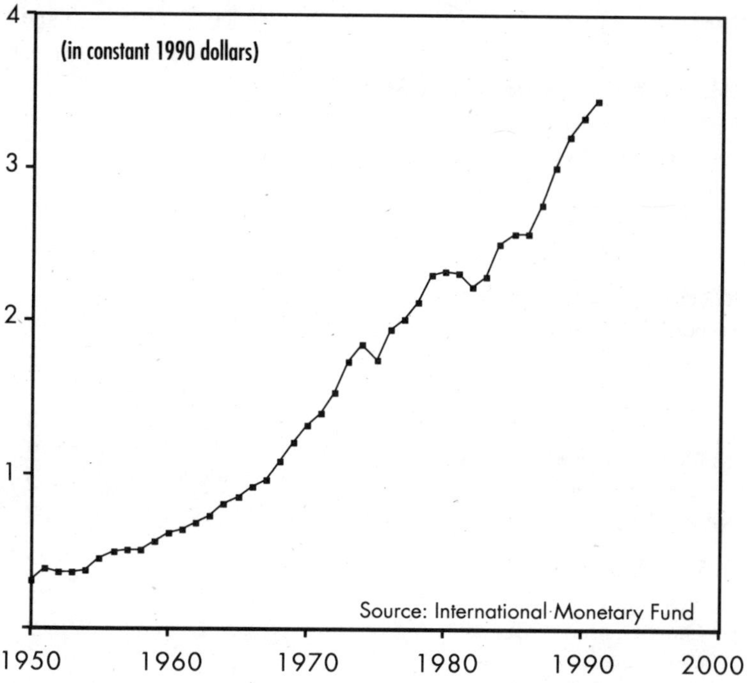

which, though lauded by business leaders, could pose a threat of sharply increased traffic and pollution. Meanwhile, heated debate over the Maastricht Treaty, the next step toward a European union, is continuing both in countries that are already EC members and in those that are considering joining. Under the treaty, more powers will be ceded from national governments to the Community—including more powers to make and enforce environmental policies.[4]

On the other side of the Atlantic, the recently completed negotiations on the North American Free Trade Agreement (NAFTA) between Canada, Mexico, and the United States have led to an agreement that will gradually phase out tariffs and reduce restrictions on investment if it comes into effect as

TABLE 1.

Value of World Trade, With Breakdown of Trade in Goods, 1990

Trade	Value (billion dollars)	Share (percent)
World Trade in Goods and Services		
Primary Products	927	22
Manufactured Products	2,445	57
Other Goods[1]	113	3
Commercial Services[2]	810	19
Total World Trade[3]	4,295	100
World Trade in Goods		
Primary Products		
Food	329	9
Fuels	369	11
Raw Materials	105	3
Nonferrous Metals	71	2
Ores and Minerals	53	2
Subtotal[3]	927	27
Manufactured Products		
Machinery and Transport	1,237	35
Other Consumer Goods	312	9
Chemicals	298	9
Other Semimanufactures	264	8
Clothing	113	3
Textiles	111	3
Iron and Steel	109	3
Subtotal[3]	2,445	70
Other Goods[1]	113	3
Total Trade in Goods[3]	3,485	100

[1]Includes gold, arms and ammunition, and other goods not classified elsewhere.

[2]Service data are less precise than merchandise data and are likely to be understated.

[3]Columns may not add up to totals due to rounding.

SOURCE: General Agreement on Tariffs and Trade, *International Trade 90-91, Vol. II* (Geneva: 1992).

planned at the beginning of 1994, creating the world's largest trading bloc. But NAFTA will need substantial modifications, through side agreements or other means, if it is to protect national environmental prerogatives while using the lever of trade concessions to ratchet up the level of environmental protection throughout North America.[5]

Reconciling the demands of global commerce with the equally compelling needs of a troubled biosphere will not be easy. Numerous short-term conflicts pit free-trade interests against environmental imperatives; yet in the long run, the interests of both coincide.

In the past half-century, governments have pursued freer trade among nations on the theory that the gains to global prosperity would benefit all, outweighing any temporary losses to groups or individuals. But as the 21st century approaches, it is now clear that the greater economic prosperity to be gained from unrestrained free trade is no longer justifiable—or even possible—as an end in itself. As reflected by their universal acceptance of the principle of sustainable development at the Earth Summit in June, 1992, the world's governments are coming to recognize that when economic development depletes the resource base upon which future productivity depends, it is counterproductive—and violates our obligations to future generations.[6]

Trade can bring greater prosperity and improved quality of life, if properly managed, but if not it can become an engine of enormous destruction. Now that policymakers have accepted the need for trade to promote sustainable development, they face the formidable task of determining how the rules of trade can be revised to help achieve that aim.[7]

The Commodity Trade

More than a quarter of the world's physical trade involves goods derived directly from the natural resource base that underpins the global economy—"primary products" such as timber, fish, and copper. For a number of these commodities, exports constitute a significant share of total production—about a third of all plywood, and nearly half of the minerals used to produce aluminum, for example. (See Table 2.) Trade therefore exerts a strong influence on the health of the global resource base. And for particular countries, the impact of trade is even greater—both in its economic value and in its environmental cost. Iceland, for example, exports half of its fish catch; yet its fishing fleet, which yields almost half of its export revenue, is rapidly depleting the very resource on which its economy depends. (See Table 3.)[8]

The industrial countries account for much of the commerce in primary products, but it is the developing countries that are most dependent on such trade—and most vulnerable to its environmental costs. The developing countries are net exporters of food, raw materials, minerals, and fuels to the industrial world, and their primary products are more likely to dominate total exports. More than 98 percent of the total exports of Bolivia, Ethiopia, Ghana, and Nigeria, for instance, are primary products—compared with 24 percent of U.S. exports and only 2 percent of Japan's. Many developing countries are thus particularly vulnerable to trade-inflicted damage to their natural resources, yet cannot afford to forego the foreign exchange these exports can generate—a dilemma that can only be resolved by strategies that promote more sustainable production.[9]

The tropical timber trade—both legal and illegal—has been a particular focus of international concern because of the role it plays in rain forest destruction. The value of total tropical timber exports ($6 billion in 1991) is beginning to fall as commercially valuable forests are decimated in country after country to meet consumer demands in Europe, Japan, and North America. In Nigeria, over-zealous logging has turned the export industry to sawdust over the past decade. In Thailand and the Philippines, timber exports have also declined so sharply that those countries now have to buy more wood than they sell. Several other countries, including Côte d'Ivoire and Ghana, will soon make the same transition from exporters to net-importers—poignant reminders that when a country squanders its resource base it also loses the economic benefits that base once provided.[10]

Of course, the timber trade is not the only cause of defor-

TABLE 2.

Trade in Selected Primary Products, 1990

Product	World Production (million tons)	Exports (million tons)	Exports as Share of World Production (percent)[1]
Plywood	32	11	33
Logs[2]	1,171	85	7
Coffee	6	5	84
Fish[3]	99	38	38
Grain	1,955	224	11
Beef	51	4	9
Bauxite and Alumina	112	52[4]	47
Iron Ore	980	395	40
Crude Oil	3,019	1,386	46
Natural Gas[5]	1,746	267	15
Coal[6]	3,316	358	11

[1]Percentages may differ from the data because of rounding.

[2]Technically, "industrial roundwood."

[3]1989 data.

[4]Reflects only seaborne trade but comprises the vast majority of total exports.

[5]In tons of oil equivalent (from terajoules).

[6]Also includes trace amounts of peat and wood.

SOURCE: Worldwatch Institute, based on sources cited in note 8.

estation in the tropics. Fuel wood gathering and the clearing of land for agriculture and grazing are also major causes, along with the felling of commercial timber for domestic use. Still, the timber trade is a more important factor than it might appear from numbers alone, as an incursion of commercial logging often sets in motion the destruction of a far larger area by spurring road construction, which in turn opens the way to mining, farming, and other activity.[11]

TABLE 3:

Countries Heavily Dependent on Exports of a Single Commodity[1]

Country	Commodity	Export as Share of Total Production of the Commodity[2] (percent)	Commodity's Share of Total Export Earnings (percent)
South Africa	Gold	100[3]	30
Botswana	Diamonds	100[4]	80
Burundi	Coffee	99	75
Fiji	Sugar	93	36
Costa Rica	Bananas	93	22
Saudi Arabia	Petroleum	86	87
Burkina Faso	Cotton	83	45[5]
Malawi	Tobacco	71	68
Iceland	Fish	50[3]	47
Malaysia	Logs	40[6]	11[7]
Uruguay	Beef	40	13

[1]1990 data unless otherwise indicated.

[2]By weight.

[3]1991 data.

[4]All of Botswana's diamonds are eventually exported although some may be temporarily stockpiled.

[5]1988 data.

[6]Technically, "industrial roundwood."

[7]1989 data, using IMF "logs and timber" category.

SOURCES: Worldwatch Institute, based on sources cited in note 8.

The impact of timber trade is particularly great in areas where production is heavily geared to the export market. In 1988 (the most recent year for which full data are available), just six countries accounted for more than 95 percent of the world's legal exports of tropical timber products by volume. Malaysia alone accounted for 62 percent of such exports, and Indonesia for 22 percent. Singapore, the Philippines, Cameroon, and

Liberia combined made up 12 percent.[12]

In Malaysia, at least half the trees felled for timber are exported, earning $1.5 billion per year in foreign exchange. But this revenue has come at a devastating environmental cost to the East Malaysian states of Sarawak and Sabah, which supply Japan with more than 90 percent of its tropical imports. Environmentalists predict that in Sarawak, the rapidly dwindling forest will be gone in as little as five years. That would obliterate the homeland of the local Penan people, who are tenaciously struggling to halt the cutting.[13]

Unfortunately, in the absence of any common regulation, the imposition of restrictions in one country can lead to massive degradation elsewhere. For instance, shortly after Thailand imposed a logging ban in 1989 to save what was left of its diminishing forest cover, the Thai logging industry simply turned to neighboring Laos and Myanmar. Similarly, the restrictions on logging imposed in Thailand and Vietnam ignited an explosion of logging—both legal and illegal—in war-torn Cambodia. According to a United Nations Development Program (UNDP) report, total exports were expected to reach nearly 1.2 million cubic meters in 1992, more than four times the amount of harvesting that could be sustained. Nearly 2 million hectares of forest—more than one-fifth of the country's total hardwood growth—has been destroyed over the last five years, according to satellite monitoring. Authorities recently imposed a ban on log exports, but it is evidently being widely ignored. If nothing is done to stop the booming trade, one U.N. official on the scene projects that "in 5 or 10 years there will be nothing left."[14]

Among the categories of trade likely to be most profoundly affected by ongoing trade negotiations is that of agricultural food commodities, which includes such varied products as grains, coffee, rubber, and cattle, and accounts for 10 percent of the value of world trade. Agricultural trade barriers, including subsidies and import restrictions, are among the targets of both the Uruguay Round of GATT, and the NAFTA. The far-reaching changes sought by these agreements have profound environmental and social implications—to which the negotiators have so far given only cursory attention.[15]

Little has been done, for example, to weigh the effects of these changes on local agriculture, particularly in developing countries. Yet one of the changes mandated by NAFTA, the reduction of barriers to selling subsidized corn from the U.S. Midwest, is likely to put millions of Mexican peasant farmers out of business. They may then have little choice but to migrate into overcrowded, polluted cities—or to head for the U.S. border. Similarly, if the Uruguay Round of GATT goes through, many more developing countries will be required to remove agricultural import barriers. The net effect may be to destroy the capability of traditional, local agriculture to remain competitive—an especially grievous loss not only because uprooted farmers rarely have any viable alternatives, but also because native agriculture can be more ecologically attuned and biologically diverse than the monoculture crops that replace it through large-scale trade. In Yemen, for instance, millennia-old sustainable low-rainfall agricultural practices were abandoned after import barriers were removed in the early 1970s as a precondition to receiving U.S. food aid.[16]

On the other hand, the Uruguay Round may somewhat counterbalance these effects by also reducing agricultural subsidies—thereby cutting back the overproduction in industrial countries that is partially to blame for world markets being flooded with cheap grain. This could raise the price that farmers in developing countries receive, helping to increase production and self-sufficiency.[17]

The economic interests of some local farmers may be overwhelmed, however, by the large-scale interests of national agricultural economies. According to the UNDP, developing countries lose some $100 billion in agricultural sales per year as a result of quotas, tariffs, and other trade barriers imposed by industrial nations. The new GATT rules will likely improve this situation somewhat by reducing tariffs on such Third World agricultural products as tea, coffee, pepper, oil palms, cassava, and natural rubber. This will create jobs and generate revenues for some of the countries that have suffered most from the North-South trade imbalance—and could conceivably make it more feasible for these countries to invest in environmental sustainability. But the removal of these barriers in the absence of careful policy reforms could also have negative environmental and social effects, such as the expansion of agricultural land into the rain forest and the promotion of cash crops over subsistence agriculture. Women and the poor are often the victims of such shifts because of their lack of access to land and other resources needed to participate in cash crop agriculture.[18]

While reducing agricultural subsidies in the industrialized countries may not fully counterbalance the threat to small-scale farming resulting from removal of trade barriers, it could have another broad environmental benefit. Reducing subsidies would cut back the artificially high domestic prices that spur overproduction (and dumping of the excess on international markets at destructively *low* prices). And in reducing excess production, they would also cut back on a panoply of other excesses: in the use of pesticides and fertilizers, clearing of forests, draining of wetlands, and depletion of water supplies.[19]

Somalian exports of sheep, goats, and cattle have grown at least tenfold since 1955.

On the other hand, because some existing farm subsidies (particularly in the United States) are explicitly linked to the adoption of sound conservation practices, subsidy cuts—like cuts of trade barriers—will have to be carefully targeted lest they backfire. In the industrial countries, many small-to-medium-scale farmers benefit from subsidies, just as some of their counterparts benefit from trade barriers in the developing countries. Like their Third World counterparts, many of these farmers grow more diverse crops, better suited to sustainable agriculture than the monoculture commodities with which they must compete; yet they could be rudely put out of business if the subsidy reductions are not accompanied by other efforts to protect them.[20]

Beef, and to a lesser degree other livestock commodities, present special difficulties for the reconciliation of trade rules with environmental protection, because of their huge demand for grazing land. In Costa Rica, for instance, the lure of the export market for beef spurred a massive clearing of the rain forest for cattle ranching. During the 1960s and 1970s, when Costa Rica was exporting large amounts of beef to satisfy the booming U.S. appetite for hamburgers, the country's rain forest was reduced to just 17 percent of its original size through agriculture and grazing. In Botswana, similarly, heavy dependence on beef exports has resulted in land degradation from overgrazing. Nearly half of the country's beef production is exported, much of it bound for the European market.[21]

Even in Somalia, which starkly demonstrates the principle that starvation is a problem of food distribution more than of production, agricultural exports have been spurred at the expense of the country's long-term environmental health, according to research by ecologist Bruce Byers. Somalian exports of sheep, goats, and cattle have grown at least tenfold, and camels twentyfold, since 1955. While providing desperately-needed revenue (livestock accounted for two-thirds of the country's export earnings in 1987, and 42 percent in 1989), this has also contributed to a tragic breakdown of the traditional, ecologically-sensitive nomadic system of livestock rearing—leading to overgrazing, soil erosion, and the degradation of rangelands, all of which diminish the ability of the land to provide sustenance for the Somali people.[22]

Along with forests, fields, and rangelands, even the oceans are being stretched to their ecological limits by the stimulus of trade. Studies of the oceans' carrying capacity suggest that the global production of fish may already have reached the maximum "take" that can be sustained. Overfishing now threatens most major commercial fishing grounds, and a principal factor is the lure of trade. About 38 percent of all fish caught worldwide are exported. In Iceland, which relies on fish exports for more than half its foreign exchange, a 40-percent quota cut thought necessary to restore the health of the local cod fishery would have caused the country's gross national product (GNP) to decline by as much as 4 to 5 percent. Faced with this Damoclean choice, the government decided to cut the quota by only 27 percent for 1993—unavoidably imperiling the future of its most important economic asset.[23]

As fish stocks are depleted in industrial-country waters, and prices rise as a consequence, northern consumers are turning increasingly to developing countries to satisfy their demand—which is continuing to grow rapidly despite the rising prices. Between 1974 and 1987 the volume of fish exported from developing nations increased nearly fourfold. With growing numbers of Third World fisheries now becoming depleted as well, overfishing for export markets often means depriving small-scale fishers of their bounty. It also drives up the price of domestically-available fish to the point where it is beyond the means of most local people. In Senegal, for instance, species once commonly eaten throughout the country are now either exported or available only to the elite.[24]

In Senegal, fish species once commonly eaten throughout the country are now either exported or available only to the elite.

Shrimp, like most other kinds of seafood or fish, is in growing demand throughout the world—with disturbing environmental consequences. In Asia and Latin America, shrimp farming for the export market has led to the clearing of large areas of coastal mangrove swamps. Once regarded as wastelands, mangrove ecosystems are now recognized as playing a critical role in protecting coastlines and serving as spawning grounds for oceanic fisheries, as well as in providing sustenance for local people. Yet, in the Philippines, mangroves are projected to be completely destroyed by prawn farming within a decade.[25]

It is not only the surging demand for biological commodities, however, that is eroding the Earth's biological capital. Mining and mineral processing have enormous environmental costs that have rarely been included in commodity prices, ranging from the destruction of huge tracts of land to the generation of prodigious quantities of pollution and waste. Developing countries often rely heavily on mining to earn foreign exchange, but for the people who gain this income, it is often a Faustian bargain. A single copper mine in Papua New Guinea, for example, yielded 40 percent of the country's export earnings, but it also dumped 130,000 tons of metal-contaminated tailings into the Kawerong River each day. The mine was finally closed in 1989, but only after a civil war—ignited at least in part by local citizens' anger over the environmental damage from the mine—forced the government from power.[26]

Developing countries are also the sites of most *unexploited* mineral reserves, which suggests that these nations will account for a mounting share of the total damage from mineral extraction in the years ahead. The Third World is also carrying a growing share of the burden of pollution-intensive and energy-intensive metal processing. Refined copper production grew by 33 percent during the 1980s in developing countries, as compared with only 13 percent in the industrial world. The share of world aluminum production centered in developing countries increased from 14 percent in 1980 to 35 percent in 1990, and is projected to climb to 44 percent by 1995.[27]

Finally, of all the major commodities in the global marketplace, perhaps none pose more ominous environmental threats than fossil fuel energy supplies. In 1990, energy exports—mainly oil, gas, coal, and transmitted electricity—totalled 3.5 billion tons of coal-equivalent. That amounted to one-third of the total energy produced worldwide.[28]

Some energy trade offers environmental benefits. For instance, burning natural gas produces less air pollution than burning coal or oil, so transporting it across national borders by pipeline allows gas to be widely substituted for more polluting alternatives. The Organisation for Economic Co-operation and Development (OECD) estimates that using gas (much of it from northern Africa and the Commonwealth of Independent States) instead of coal in 20 percent of Western Europe's power plants would cut carbon emissions from the utility sector by 8 percent, sulfur dioxide emissions by 20 percent, and nitrogen oxides by 8 percent.[29]

It should be noted, however, that such benefits accrue only

to the importing country. With fossil fuels, as with fresh-cut trees, the importer gains the economic benefit of using the commodity without suffering many of the environmental consequences of its extraction. For example, a dam built on the La Grande River in northern Quebec, in part to generate electricity for export to New England, flooded an area half the size of Belgium. The water leached naturally occurring mercury from the flooded land, poisoning the fishing grounds of the indigenous Cree Indians. Similarly, the Soviet Union for many years engaged in a process of environmental and economic self-liquidation by supplying large quantities of oil and gas to Eastern and Western Europe. This provided a key source of foreign exchange, but also left a ruinous environmental legacy. Only recently, after the political dissolution that followed, has the extent of the destruction in the producing regions begun to come to light.[30]

Halting the environmental toll of the global commodity trade may require widespread implementation of at least four strategies: reforming the practice of "tariff escalation" that has discouraged developing countries from acquiring processing industries; diversifying the industrial base of developing countries to include less environmentally costly kinds of commerce; incorporating "environmental missions" into existing commodity organizations to focus attention on the interdependence between economic and environmental health in their industries; and providing debt relief to those developing countries that have become self-destructively export-dependent.

Stemming tariff escalation (which tends to confine developing countries to the early stages of production lifecycles) would take pressure off the natural resource base, since profits made from processing tend to be less environmentally damaging than those from extraction. For instance, a study by the International Institute for Applied Systems Analysis (IIASA) in Laxenburg, Austria found that mining bauxite to make alumina uses about 10 percent of the total energy used in making the final aluminum product, and produces 90 percent of the wastes while accounting for only 10 percent of the profits. The second stage, processing alumina into raw aluminum, uses about 80 percent of the energy and produces 9 percent of the wastes, while accounting for 20 percent of the profits. In contrast, the final stage of manufacturing refined metal goods generates 70 percent of all profits while requiring only 10 percent of the total energy use and producing only 1 percent of the wastes.[31]

Tariff escalation might be described as one of the controlling mechanisms of the global industrial hierarchy, whereby industrial countries impose higher tariffs on more-refined (value-added) goods than on less-refined or unrefined (primary) ones in order to protect their own processing industries. For instance, Japan levies a tariff of 9 percent on plywood imports, but allows logs to enter duty-free. Similar patterns control the movement of processed versus "raw" forms of cocoa, fish, minerals, and rubber. If the barriers were removed, developing countries would more often find it profitable to acquire their own processing industries—which in turn would make it easier for them to protect their resource bases. The Uruguay Round is expected to make some inroads into this blatantly protectionist practice, though it will not eliminate it altogether.[32]

Diversifying the industrial base offers several means of improving environmental protections by *strengthening*, rather than weakening, Third World economies. One strategy is for developing countries to move away from heavy reliance on natural resource extraction by developing more light manufacturing industries—as newly industrialized nations like Taiwan and Korea have so successfully done with textiles, for example. Tariff reductions on these products, contemplated under both the Uruguay Round and NAFTA, would help accelerate this process. The trouble with this strategy is that it appears to benefit developing countries at the expense of affected workers in the (mainly industrial) countries whose tariffs are being reduced.[33]

Another approach is to pursue alternative types of trade that are not as environmentally or socially destructive as conventional exports. For example, one way to preserve forests while providing a livelihood to those who live in them is through trade in non-timber forest products such as spices, rubber, Brazil nuts, and rattan (palm stems used for wicker furniture and baskets). The revenues can be substantial: Southeast Asia's rattan trade is worth more than $3 billion a year, and palm hearts and Brazil nut exports netted $20 million for Brazil in 1989. While there is no guarantee that alternative forest products will be harvested sustainably, experience shows that they are more likely to be than timber is.[34]

The pioneers in this effort are Brazil's rubber tappers, who have worked to attain rights to "extractive reserves" where these non-timber forest products can be produced on a sustainable basis and sold to companies like Ben and Jerry's ice-cream and the Body Shop, which derive environmental cachet for their products by their association with this cause. Extractive reserves have also been established in Peru, Guatemala, and elsewhere, and the same strategy for boosting sustainable exports could be expanded more widely.[35]

The final stage of manufacturing refined metal goods generates 70 percent of all profits while producing 1 percent of the wastes.

In a similar spirit, the "fair trade" movement seeks to address social concerns about trade by promoting trade in goods that conform to certain social criteria, including adequate working conditions and higher prices paid to producers, both to compensate for stagnant commodity prices and to ensure that profits are not lost to middlemen. Environmental considerations could be added to the criteria. Products now being traded this way include coffee grown by worker cooperatives, and the handicrafts sold through catalogues by alternative trade groups like Pueblo to People. One particularly successful example is coffee sold by Holland's Havelaar Foundation, which comes with a "fair trade" mark affixed to it to certify that it was bought on fair terms. Imports of the coffee into the Netherlands have risen from 200 metric tons in 1988 to 2,800 in 1992, and the fair-trade coffee is now sold in 90 percent of the country's supermarkets. It is also sold in Belgium and Switzerland, and is soon to be launched in France and Germany. In both Europe and the United States, efforts are underway to develop a similar mark to be applied to other products.[36]

In addition, importing countries might consider granting preferential tariff treatment to sustainably produced goods, providing further incentive for producing countries to emphasize them. At the prodding of Ecover, a Belgian company that makes soap using environmentally sustainable materials and processes, the government of Columbia has requested exemption from EC import duties for oils produced from certain organically grown plants such as lemon grass. Ecover plans to buy ten different such oils from Columbia. Because farmers would earn 67 times as much per hectare from this group of products as from coffee, and 3.2 times as much as from cocaine, this would be a mutually beneficial transaction.[37]

A strategy that would involve more of the mainstream commodities trade is for commodity organizations to adopt environmental missions, as the International Tropical Timber Organization (ITTO) has done. When ITTO was created in 1985, its members, both producer and consumer nations (who now number 50)—gave it the explicit task of seeing that trade in tropical timber shifted to a more sustainable footing. In 1990, they set the year 2000 as a target date by which all trade in tropical timber should be from sustainably managed stands.[38]

So far, progress toward this goal has been minimal, with both producer and consumer nations seeming to lack the political will necessary to find solutions. But the members have not given up, and are now debating whether the agreement should be expanded to include wood from temperate forests, which are also being decimated in some areas. To turn rhetoric to action, the ITTO will need to institute real policy changes, such as labeling requirements to bring consumer pressure to bear, and technical assistance in developing more sustainable practices for producer countries. Other commodity groups that at least pay lip service to environmental missions include the International Jute Organization and the International Sugar Organization.[39]

Of course, trade organizations would—by definition—need to have a clear economic rationale for such a mission. The ineffectiveness of the ITTO mission may also be due in part to the vagueness of the sustainable footing goal. A more successful approach might be one that simultaneously and pointedly addresses a specific *economic* concern—the problem of stagnant commodity prices—along with the environmental problem of over-exploited natural resources.

A technique called "mutual tariffs" might accomplish this goal through the intervention of commodity organizations. Under such a program, exporters of primary commodities would impose an environmental export tariff on their goods, the proceeds of which would be funneled to sustenance of their resource base. Importers could require the imposition of such a tariff (by the exporter) as a condition of market access, in order to prevent a country that adopted one from being unfairly disadvantaged—and to remove any incentive for that importer's competitors not to adopt environmental tariffs of their own. Widespread use of such tariffs would be an acknowledgement that chronically low prices for commodities are of environmental concern, since under existing price structures many countries are forced to produce more and more in order to generate the foreign exchange they need to pay back debts or import goods. In essence, it would be one way of ensuring that prices for commodities reflect their full environmental costs.[40]

Ultimately, the environmental destruction driven by the demands of burgeoning trade can't be fully halted without debt relief.

In the long run, however, stemming the environmental toll of the commodities trade will require not only more sustainable pricing and resource management by the producers themselves, but more far-reaching changes in overall global development strategies. Ultimately, the environmental destruction driven by the demands of burgeoning trade probably can't be fully halted without major debt relief for developing countries. This may be essential to reducing the incessant pressure to liquidate natural assets (via escalating export-oriented production) in order to repay foreign banks. Allowing debt-strapped Third World producers to escape this debilitating treadmill would facilitate the kinds of small-scale, grassroots development efforts that are capable of reducing both poverty *and* environmental degradation in developing countries.

In one recent precedent, a U.S.-backed program to develop free trade throughout the Americas, the Enterprise for the Americas Initiative, recognized the link between debt and trade by including $777 million in debt relief. On a much larger scale, as the World Bank and the International Monetary Fund proceed with their agenda of restoring creditworthiness to debt-ridden Third World countries through "structural adjustment" programs, it is critically important that their programs explicitly incorporate environmental concerns. These programs would then become powerful tools for encouraging policy reforms such as environmental taxation. In doing so, they would help to discourage over-exploitation of resources, instead of actually exacerbating the problem through single-minded devotion to boosting exports at any cost.[41]

Bad Goods and Green Goods

While much of the environmental cost of trade is incurred at the "extraction" end of a product life cycle—far from the eyes of affluent industrial-country consumers—there are other kinds of damage that are more apparent in the nature of the products or byproducts themselves.

Some widely-traded products are more notable for the inherent environmental significance—positive or negative—of the products themselves, than for the effects of producing them in large quantities for export. To some degree, the overall "environmental balance of trade" in the coming generation will be determined by the extent to which the mix of traded products can be managed to minimize the movement of hazardous products or endangered animals, while encouraging those that have environmental as well as economic benefits.

Of the former, the most notorious are the waste products of

industrial society—chemical and otherwise. World attention was drawn to the "toxic trade" in 1988, when the dumping of hazardous waste in a Nigerian village by a European trader and a round-the-world tour by a "garbage barge" toting incinerator ash from Philadelphia aroused widespread outrage. Since then, the trade has been less well publicized, but hardly diminished. According to Greenpeace, at least 10 million tons of waste of all types has been exported over the last several years, more than half of it going to Eastern Europe or developing countries, where regulation has tended to be lax.[42]

In 1989, the international community endorsed the Basel Convention on waste export, which regulates this trade by requiring exporters to notify the recipient nation of a shipment and to receive approval for it before proceeding. After a three-year wait, the treaty came into effect in May 1992 when it received the necessary 20 ratifications. Many developing countries found this little cause for celebration, however, contending that the Basel accord legitimizes a trade that should be banned outright. At a recent meeting of the treaty members, developing countries called for the treaty to be toughened. However, neither the United States nor any of the European Community countries except France has yet ratified even the original convention.[43]

Trade agreements could make it still more difficult to control the exporting of waste. The removal of border controls in Europe, and the increased traffic across North American and other international borders, could overwhelm efforts to control illicit traffic. Moreover, these agreements could dictate that waste is a product not unlike others in international trade, making it difficult for nations to unilaterally ban exports or imports.[44]

In the European Community, this concern is more than hypothetical. Under pressure from Dutch waste exporters, the European Commission recently challenged a ban on hazardous waste imports passed in the Belgian province of Wallonia, in response to a public uproar over a waste import scandal. In May 1992, the European Court of Justice issued a somewhat equivocal ruling prohibiting blanket bans on hazardous waste imports under existing EC law, yet allowing for restrictions on specific shipments. Notably, the ruling did support the principle that free trade should give some ground to environmental concerns in this case, on the grounds that waste is not an ordinary good and should be disposed of as near its source as possible. EC waste trade legislation agreed to in October 1992 supports this interpretation, thereby protecting Wallonia's law, as well as waste import bans being considered by France, Portugal, and the United Kingdom.[45]

Of course, numerous other dangerous products are disseminated through trade—sometimes *because* of their hazardous properties. In Canada, for instance, when domestic sales of asbestos declined due to public health concerns, the industry collaborated with the government to promote sales abroad. Ninety-five percent of the asbestos Canada produces is now exported—more than half of it to the Third World. In years past, the government offered free samples to targeted countries such as Thailand and India, as an enticement. The Canadian industry has even gone so far as to challenge in U.S. courts the *U.S.* ban on asbestos use imposed in 1989. The Canadians' concern was that Third World countries might model their legislation on the U.S. law, thereby keeping out Canadian asbestos. The court ruled in Canada's favor, but not on trade grounds; it averred that GATT would be the proper forum to consider those questions.[46]

Pesticides producers have pursued a similar strategy of international expansion, as a means of perpetuating sales of products found too dangerous for domestic use. The U.S. General Accounting Office (GAO) estimated in 1989 that 25 percent of the pesticides exported from the United States were unregistered or severely restricted for domestic use, due either to their harmful qualities or to the lack of a domestic market for them. In one of the more egregious cases, in 1991 the Velsicol Chemical Corporation exported—with no restrictions on use—681 tons of the pesticides chlordane and heptachlor, probable human carcinogens which the U.S. Environmental Protection Agency has prohibited for virtually all purposes. Banned pesticides sometimes return on imported food, a phenomenon known as the "circle of poison."[47]

UNEP and the U.N. Food and Agriculture Organization (FAO)

have created a voluntary system whereby developing countries will be notified of actions taken by industrial countries to ban or restrict the use of certain pesticides or chemicals. The recipient nations are then given the option of rejecting shipments. In addition, a working group of the GATT several years ago examined the question of "domestically prohibited goods" with the aim of establishing a system of informed consent similar to—but farther-reaching than—the UNEP and FAO guidelines. However, the talks broke down in May 1991 when the United States objected to the draft partly because the GATT effort was perceived to be more binding than the UNEP and FAO system. The talks have not resumed since, though several developing countries are pushing for them to be reconvened.[48]

Another category of trade in which great environmental harm is represented by the products themselves—though it is harm of a very different kind—is the traffic in wildlife and wildlife parts. Though habitat loss is the world's leading cause of diminishing biological diversity, there are some species—such as the Orinoco crocodile and the Sumatran rhinoceros—that are particularly prized on the international market and for which trade is a leading threat. For others, it may be the *last* threat. To a population already decimated by habitat loss and other forces, even moderate pressure from commercial trade may be enough to bring about extinction.[49]

To a population already decimated by habitat loss and other forces, pressure from commercial trade may be enough to bring about extinction.

Global wildlife trade is valued at $5 to $8 billion annually, some 30 percent of which is illegal. Each year, 30,000 monkeys and other primates are shipped across international borders, along with 20 to 30 million pelts, 500,000 parrots, 400 to 500 million ornamental fish, 1,000 to 2,000 raw tons of corals, 7 to 10 million cacti, and 1 to 2 million orchids. Nations have had some success, through the Convention on International Trade in Endangered Species of Wild Fauna and Flora (CITES), in controlling trade in products from such well-known species as the African elephant and the snow leopard. However, trade in many other threatened species, such as the black rhinoceros, continues apace.[50]

Regional trade pacts may exacerbate the difficulties in controlling this traffic. For instance, a European Community task force created to study the environmental impact of the single market warned that the removal of border controls may impede efforts to monitor wildlife trade. In North America, a study by the World Wildlife Fund projects that the increased tourism and transport likely to result from NAFTA—along with the tariff reductions it mandates—will lead to more trade, both legal and illicit, in furs, exotic leather goods, parrots, stuffed birds of prey, sea turtle products, and other valuable and endangered species.[51]

For all the environmental pressures increased by the expanding of international markets, however, there are also a number of benefits. Global markets disseminate products and technologies that can help speed the adoption of cleaner, more energy- and materials-efficient industrial processes throughout the world. As environmental laws are strengthened in developing countries, sales of pollution control equipment are surging. Japan, the United States, and Europe (mainly France, Germany, and the United Kingdom) collectively exported $20 billion worth of such equipment in 1990—about 10 percent of the industry's total production. Renewable energy industries, similarly, are aiming at international markets. In one particularly good year, for instance, wind turbine manufacturers exported over 550 megawatts' worth of equipment, equivalent to about one-half the generating capacity of a typical large nuclear power plant, and valued at $720 million. In 1992, $200 million worth of photovoltaic cells—more than half of the world's total production—were exported by their manufacturers.[52]

International trade also can put pressure on companies to match the environmental innovations of their foreign competitors, as in the U.S. motor vehicle industry's response to Japan's advances in fuel efficiency. Other examples include the early success of Swedish industries in developing low-flow toilets, which have now been replicated around the world, and a U.S. lead in some air pollution control technologies that has now been overtaken by the Japanese and the Germans.[53]

Provisions to reduce restrictions on environmental services, such as consulting and engineering, are included in both the NAFTA and the latest draft of the Uruguay Round. And by reducing tariffs and other trade restrictions, these agreements can help to facilitate trade in greener goods. In addition, mechanisms to support the transfer of environmental technologies could be included in a side agreement on the environment, to supplement the NAFTA.[54]

Pollution Havens?

The free trade agenda has become as much concerned with free *investment* as with trade itself. One of the abiding goals of the NAFTA, Europe's single market, and the Uruguay Round is to free restrictions on foreign investment, making it easier for multinational corporations to strategically locate various facets of production wherever costs are lowest around the world.

The trend toward greater mobility of capital is of environmental concern for two important reasons. First, there is a danger that capital will flow to countries with lax environmental laws or enforcement, creating "pollution havens." Second, there is the fear that the industries which do operate in such havens will have an ill-gotten advantage over those based in countries with stringent environmental laws. This could be a double-barrelled disincentive to the adoption of progressive environmental laws—as in the recent case of the European Community's reluctance to adopt a proposed tax on carbon dioxide emissions in the absence of a similar commitment by its industrial competitors, Japan and the United States.[55]

Despite the dampening effect of tariff escalation (as noted earlier) on the development of high value-added processing in developing countries, some manufacturing industries—including textiles, leather, iron and steel production, and chemicals—

are increasingly concentrating in these countries. They accounted for only 19 percent of global textiles production in 1975, for instance, but for 26 percent by 1991. Their share of iron and steel production doubled over the same period, climbing to 20 percent. Altogether, developing countries' share of world exports of manufactured goods increased from 4 to 19 percent between 1955 and 1989. Of course, these numbers mask great variation in individual circumstances. The newly industrialized countries of Southeast Asia have been rapidly expanding exports of everything from shoes to cars, while much of the developing world remains highly dependent on trade in such primary goods as coffee and timber.[56]

The growth of manufacturing in developing countries has a number of explanations. To begin with, development planners—doubtless influenced by the history of the industrial revolution in the North—have touted manufacturing as the route to prosperity in these countries. Their strategy has been abetted by the availability of cheaper labor or energy, accessibility of raw materials, and the industrial world's increasing shift from heavy manufacturing to high-tech and service industries.

The extent to which environmental regulations in the industrial countries might be quickening these trends has been a subject of debate for years. Most studies suggest that pollution control expenses alone are generally not a large enough share of total costs to damage competitiveness or make it worthwhile for a company to relocate. The U.S. Census Bureau estimates that these costs average just over 1 percent of value added for more than 400 industries. However, they are significantly higher for some, including copper smelting at 22 percent, oil refining at 9 percent, and steel production, at 7 percent. And the 1991 IIASA study found that a state-of-the-art Austrian paper mill has "pollution avoidance costs" nearly as high as the plant's raw materials costs.[57]

Studies also indicate that in certain instances environmental rules can be a decisive factor in industry location decisions. In a 1988 book, H. Jeffrey Leonard documented several cases in which particularly hazardous operations relocated in order to escape strict environmental standards. They included producers of asbestos, benzidine dyes, and a few pesticides, along with some mineral processors. He also found that some countries—including Ireland, Mexico, Romania, and Spain—deliberately set themselves up as pollution havens in the past, though they have since backed away from this policy in the face of growing public concern over the environment.[58]

More recently, researchers found that in Mexicali, along the U.S.-Mexico border, more than a quarter of the factory operators said that Mexico's lax environmental enforcement influenced their decision to locate there. A GAO study found that a sizable number of Los Angeles furniture manufacturers relocated to Mexico between 1988 and 1990—as well as to neighboring states. More than three-quarters of them cited California's more stringent air pollution regulations as a major factor in their decision to move.[59]

More than half of the 100 largest U.S. companies operate assembly plants in Mexico.

Whether attracted by low wages, lax enforcement of human rights and labor laws, or look-the-other-way environmental standards, it is clear that manufacturing—some of it hazardous—is on the rise in much of the Third World. Some of the manufacturing in developing countries is foreign-owned and some is not, but in either case there is a danger that due to aging technology or equipment and inadequate environmental protection in many developing countries, this investment causes considerable harm.

With the debate over the NAFTA, these problems have brought particular notoriety to the border between the United States and Mexico. This border region has become the site of nearly 2,000 manufacturing plants known as maquiladoras—branches of companies that are allowed to import duty-free components for processing in Mexico on the condition that the final product be exported back. Almost all the plants are foreign-owned, lured there by wages as low as 7 percent of what is paid for comparable work in the United States, as well as by the preferential tariff treatment and proximity to markets. More than half of the 100 largest U.S. companies operate assembly plants in Mexico.[60]

On the books, Mexico's environmental laws are roughly comparable and in some cases stricter than U.S. ones. But enforcement has been lax; an official of the Mexican environment ministry estimated in 1991 that only 35 percent of the U.S.-owned factories along the border comply with Mexican toxic waste laws. Though maquiladoras are required to return to the United States any waste they generate, compliance with this mandate is believed to be the exception rather than the rule. Even with the intensive focus on Mexico's environmental record in light of the negotiations over NAFTA, a GAO spot survey of six U.S.-owned maquiladoras recently found that none had obtained the environment permit that Mexican law requires.[61]

Investigations have revealed alarming conditions in the border area. At three-quarters of the maquiladoras sampled in 1991, the U.S. National Toxics Campaign found toxic discharges, including chemicals that cause cancer, birth defects, and brain damage, being emptied into open ditches running through settlements near the factories. The American Medical Association describes the border area as "a virtual cesspool and breeding ground for infectious diseases," in no small measure due to the fact that the population there has swelled to twice its former size in the last two decades, while sewage treatment remains practically nonexistent. Investigations by medical teams on both sides of the border have revealed alarming public health conditions, including elevated rates of hepatitis A and tuberculosis, in part because some of the pollution drifts or flows back and forth across the international boundary.[62]

Though the maquiladoras have received more incriminating publicity than other such sites, they are far from unique. Environmentalists in Southeast Asia, for example, have sharply criticized Japanese industries for allegedly locating extremely harmful processes abroad because they can no longer pass environmental muster at home. A Malaysian subsidiary of the Mitsubishi Kasei Corporation was recently forced by court order to close after years of protests by local residents that the plant's dumping of radioactive thorium was to blame for unusually high leukemia rates in the region. Several multinational corporations operating in South Africa, including local subsidiaries of the German-based Bayer pharmaceuticals concern and the U.S.- based Duracell battery company, have been implicated by local environmentalists in toxic catastrophes that they believe caused severe health problems among workers.[63]

Worldwide, there are at least several hundred maquiladora-like free trade zones where foreign investors are given special breaks on tax, investment, and other enticements. And there have been widespread allegations that employers in these zones violate international labor norms with impunity. For instance, an analysis by the U.S. Department of Labor found that in five of eleven nations studied, labor rights in export processing zones were more restricted than in the rest of the country. Further analysis is needed to determine if environmental laws are similarly disregarded.[64]

Environmentalists have criticized Japanese industries for locating harmful processes abroad because they can no longer pass muster at home.

Though the growth of foreign investment poses some environmental threats, it also brings some potential gains. To begin with, investment serves as a means of disseminating environmentally-friendly technologies around the world. For example, General Electric recently entered into a joint venture with the Tungstrum Corporation to produce energy-efficient compact fluorescent light bulbs in Hungary—offering environmental benefits that would not have been possible had Eastern Europe remained closed to trade with the West. A World Bank study compared the rates at which 60 different countries were adopting a cleaner process for making pulp for paper, and concluded that the new technology made its way to countries open to foreign investment and trade far more rapidly than to those closed to it. In addition, there is some limited evidence that countries which are closed to investment are more likely to pursue environmentally harmful forms of industrial development: according to another World Bank study of 25 Latin American countries, dirty industries developed faster in relatively closed economies than in open ones.[65]

The authors of these studies suggest several possible explanations for these trends. First, closed economies might protect capital-intensive, pollution-intensive industries in situations where low-cost labor would otherwise have attracted less-polluting industries. Second, industries trying to sell their goods in industrial-country markets need to please the growing numbers of "green" consumers in those markets. Finally, the equipment used by multinationals is generally newer, and therefore cleaner, than that employed by locally owned firms.[66]

A number of steps can be taken to prevent the creation of pollution havens, and any unfair trade advantages to be gained through lax environmental laws. In the debate over the North American Free Trade Agreement, Mexico promised to tighten enforcement and explicitly ruled out becoming a pollution haven. One innovative feature of the agreement is its explicit stipulation that it is "inappropriate" to relax environmental standards or enforcement in order to encourage investment. Unfortunately, noteworthy as this may be, it is—in its present form—largely exhortative. Environmentalists are lobbying to strengthen the provision either by requiring new U.S. investors in Mexico to meet U.S. laws (and imposing penalties for violators), or by imposing countervailing duties or other sanctions to compensate for shortfalls.[67]

In essence, this would be a means of linking trade concessions, or participation in free trade agreements, to respect for internationally recognized environmental standards. There is some precedent for this in the labor arena. For instance the International Labour Organization has promulgated hundreds of regulations on such issues as workplace safety and the right to organize. Under existing U.S. laws passed in the 1980s, trade, aid, and investment guarantee privileges are contingent upon respect for these internationally recognized worker rights and labor standards. Faced with the loss of these special privileges, several countries have made concerted efforts to remedy the shortcomings in their laws. In a related measure, the 1988 U.S.

Omnibus Trade and Competitiveness Act stipulated that failure to respect certain worker rights was an unfair trade practice that could lead to trade retaliation. So far, however, no complaints have been lodged. The United States has tried to have these unilateral laws endorsed by the GATT, but they have encountered widespread opposition, particularly from developing countries who see them as a violation of their sovereignty.[68]

One problem with applying this approach in the environmental arena is the lack of obvious international standards to point toward. However, the European Community has addressed this problem by promulgating a body of environmental laws governing production processes. These "directives," justified largely on trade grounds, are minimum standards that member states are free to exceed if they like. Under the NAFTA, a number of groups, including the Washington-based National Wildlife Federation and Institute for International Economics, have endorsed the idea that the "harmonization" of some production standards should be undertaken in connection with the agreement, perhaps through the intervention of the proposed North American Commission on the Environment. This would mean that a pulp and paper mill, for example, might have to meet a minimum effluent level in all three countries. While harmonization is not yet under active consideration at the international level, it could be argued that the 170 international environmental treaties already on the books represent a first step in the direction of global environmental lawmaking.[69]

Those trying to sell their goods in industrial-country markets need to please growing numbers of "green" consumers.

Some analysts argue that harmonizing environmental standards around the world is not worth attempting, as countries value environmental quality differently and have varying levels of "assimilative capacity," which is economists' jargon for unsullied air, water, and other natural resources. Yet this argument disregards the extent to which one country's problems can affect other countries' environmental quality, given the propensity for pollution to cross borders on wind or water—or through trade itself. It also disregards the desirability of creating a "level playing field" for business.

The argument against global environmental standards also betrays a misconception that weak environmental regulations are a legitimate source of comparative advantage, rather than a form of exploitation similar to that of flouting internationally recognized human rights and labor standards. While complete harmonization of environmental policies is not practical, and is not sought in any existing trade agreements, moving toward environmental parity around the world would be an important step toward minimizing some of the dangers of trade.

In the absence of an international standard, one way to move toward such parity without imposing it directly would be to permit countries to impose compensating border tariffs. This would ensure that they are not penalized in international markets for internalizing environmental costs more than their trading partners do, since lower standards do add up to a sizable hidden subsidy. I. Walter and J.H. Loudon estimate that developing countries exporting to the OECD countries in 1980 would have incurred pollution control costs of at least $5.5 billion if they had been required to meet the requirements then prevailing in the United States.[70]

For the time being, the use of such adjustments for environmental or other purposes is discouraged by OECD guidelines and would likely be overruled by the GATT, though many environmental groups favor changing these rules. There is some historical precedent for this in the area of labor and workplace standards, as in regulations imposed by Britain's Board of Trade in 1925 that provided for imposing duties on imports to compensate for lower wages and longer workdays. Developing countries are strenuously opposed to applying this concept to the environmental arena, fearing that environmental concerns will be used as an excuse to discriminate against their products. One way to help allay their concerns might be to channel the proceeds from these duties into upgrading the environmental standards of those countries which fail to measure up.[71]

A Question of Scale

When trade increases, economic activity increases—whether the trade is in raw materials or high-tech manufactured goods. From an environmental standpoint, this spur to commerce can be either a curse or a blessing—depending on whether the policies and incentives which govern that trade can point it in a sustainable direction.

In general terms, more trade means that more goods are transported around the world, which means more energy use and pollution. Transporting the 4 billion tons of freight sent by ship worldwide in 1991 required 8.1 exajoules of energy—as much as was used by the entire economies of Brazil and Turkey combined. The 17 million tons sent by plane used 0.6 exajoules—equal to the total annual energy consumption of the Philippines. Though far more goods are sent by ship than by air, it takes 47 times as much energy to carry a ton of goods by air as it does the same distance by boat. Similarly, a cargo transported by truck requires substantially more energy than the same load hauled by rail.[72]

Trade agreements will likely encourage growth in transport. The EC task force on the single market predicted that cross-border truck traffic—and the air pollution and noise that accompany it—would grow by 30 to 50 percent with the opening of Europe's borders in 1993. Air traffic was expected to increase even more. And similar expansions are anticipated in North America. Cross-border trucking between the United States and Mexico has doubled in the last five years, reaching at least 1.8 million crossings per year—with continued growth expected—even *before* any reduction of trade barriers between the two countries. If NAFTA is implemented, the U.S. government predicts that more than 12 million trucks per year—nearly seven times the present volume—could be crossing the border in both directions by the end of the decade. In both North America and Europe, this massive increase of motor vehicle traffic could bring increases in fuel consumption, air pollution, carbon emissions, and accidents involving trucks carrying hazardous cargoes.[73]

In both regions, the increased volume of goods being shipped may be offset somewhat by a reduction in the number of trucks that have to unload their cargoes at the border and return empty to their home bases due to trade barriers against the trucking industry. Before the single market came into effect, about a third of all trucks on EC roads traveled empty for this reason. Under NAFTA, as well, trucking restrictions will be lifted. And the proliferation of regional trade pacts could mean fewer goods are transported from far-flung corners of the globe. Thus, while more trade will mean more trips, the length of the trips may be shorter—at least partially mitigating the increased energy consumption. And in both Europe and North America, governments hope to head off some of the growth in road traffic by encouraging the use of rail.[74]

Beyond its transportation requirements, more trade tends to open larger markets for goods and services, which stimulates economic growth. In the absence of mitigating policies, increased growth will lead to increased energy use, materials consumption, and consequent pollution and waste generation. Thus, the EC task force report concluded that in the absence of technological change, the economic growth likely to result from the single market would increase emissions of sulfur dioxide by 8 to 9 percent and nitrogen oxides by 12 to 14 percent by the year 2010. A study by Princeton University economists Gene Grossman and Alan Krueger found that the trade and investment liberalization envisioned under NAFTA would increase annual toxic emissions into the air, water, and land—from all industrial sources—by 10.5 million pounds in Mexico, 13 million in the United States, and 8 million in Canada. The combined total surpasses the total toxic releases in the state of New Jersey in 1990.[75]

But if mitigating policies are put in place, technological change encouraged by these policies can at least partially de-link increased pollution from increased economic output. Between 1970 and 1989, for example, energy use per unit of GNP decreased by 30 percent in OECD countries. Though this was due partially to structural changes in the economy away from manufacturing and toward services, and possibly to some transfer overseas of energy-intensive manufacturing to non-OECD countries, a key factor was improvement in the energy efficiency of industrial technologies.[76]

The technological benefits that can compensate for increased industrial output include not only the indirect effects achieved when increased energy efficiency means decreased pollution, but numerous direct reductions of emissions as well. For example, while bleached pulp production in Sweden grew by 5 percent between 1986 and 1991, pulp processing innovation enabled the industry to cut its use of environmentally harmful chlorine in the production process by 78 percent.[77]

A tax on energy will limit the environmental costs of trade by encouraging use of the least energy-intensive transport option.

A related tradeoff is that the income generated by economic growth can make investments in pollution control more likely—or at least more affordable. In an analysis of sulfur dioxide and particulate pollution levels in cities at varying stages of development around the world, Grossman and Krueger found that once countries reach a certain level of national income (about $5,000 per capita), the levels of these air pollutants, which are relatively easy to control technologically, tend to decline. The presumed explanation is that at that point governments find they have the revenue—and the political will—to invest in controls.[78]

As to how much added income might be generated for developing countries by the provisions of the Uruguay Round, estimates range from $12 billion to $90 billion. Whatever the actual figure, this is money that could—in theory—be invested in environmental programs, which in turn would provide greater long-term economic stability. But contrary to the assertions of ardent free trade advocates, in the absence of major policy changes there is no assurance that any of the increased revenue would actually be put to that purpose.[79]

To encourage economic development of a greener kind, one of the most needed reforms in all countries is to ensure that products are priced to reflect their full environmental cost. If this could be done, the growth generated by trade would become far more environmentally sound. Trade would then be a consistently efficient means of distributing resources around the world, since "internalization" of the real costs would make inefficient transport unprofitable, while enhancing the profitability of trade that saves energy or reduces pollution. No country would be able to gain a comparative advantage through degrading the environment.

One of the primary ways to achieve this cost internalization is through environmental taxation. A tax on energy will limit the environmental costs of trade by encouraging the use of the least energy-intensive transport option, and in some cases by making a homegrown product more economical than an imported one. A tax on wood from primary forests would encourage the use of plantation-produced timber or non-wood pulp, reducing the impact of the timber trade on deforestation.

Although economists prefer economic incentives such as taxation because they cost less to implement, regulations can also help internalize costs. For example, strong air quality regulations require industry to pay for pollution controls, thereby internalizing some of the costs of damage to human health and infrastructure as well as alleviating the environmental impact of manufacturing for export.

The greening of trade can also be encouraged through environmental labeling programs that help harness consumer pressure for change. Successful examples include the Blue Angel program in Germany and the Green Seal effort in the United States. The European Community has recently implemented a far-reaching program of its own. The choice of strategy—environmental taxation, regulation, or consumer pressure—will depend on what's being produced and where. But the important point is that strong domestic environmental policies around the world will be necessary if the increasing volume of trade is to be compatible with sustainable development.[80]

Environmental Trade Barriers?

If environmentalists fear that free trade will damage natural resources, free trade enthusiasts worry that stricter environmental policies around the world will cause unacceptable disruptions to commerce. They raise the specter of hard-won trade reforms being undermined as governments use import restrictions, export constraints, and subsidies to meet environmental goals. In response, environmentalists emphasize the need to selectively use trade-restrictive tools to meet important environmental goals, particularly given the global nature of the economy and the environment.

Unquestionably, in some cases environmental policies and free trade goals collide, at least in the short term. That collision raises vexing questions. Under what circumstances are environmental goals legitimate grounds for suspending the usual trade rules? Are there cases in which trade considerations should override environmental ones? And who makes these difficult decisions?

National environmental rules that put restrictions on imports are among the kinds of policies that could be threatened by freer trade. It is common practice for a country to restrict the entry of products that do not meet its domestic environmental, health, and safety laws. Otherwise, the laws' effectiveness would be quickly undermined. For example, automobile manufacturers selling to the U.S. market must make cars that meet U.S. emissions standards. Farmers growing fruits or vegetables for export must ensure that pesticide residues are within the allowable limits of the importing country. And beer companies selling abroad must make sure that their bottles are refillable, if that is the importing country's law. Current trade agreements generally protect the right of countries to hold foreign manufacturers to certain standards for products sold in their markets, so long as the same conditions apply to domestic producers and all foreign competitors are treated equally.[81]

That much is fairly straightforward. Where the agreements get sticky is in the attachment of certain qualifications, such as those which seek to assure that a given measure was motivated by genuine environmental concerns and is not just a disguised form of protectionism. Governments often suspect each other of wrapping protectionism in an environmental mantle. For example, the United States has long maintained that a European Community ban on beef produced with bovine growth hormone is really an attempt to protect EC cattle producers from U.S. competition. And in a dispute with Ontario, the United States claims that a recently imposed Ontario tax on all alcoholic beverages sold in nonrefillable containers was established not for environmental reasons but to keep out U.S. beer, which is mostly sold in cans. Similarly, Mexican citrus growers claim that a U.S. rule requiring imported limes to undergo a treatment for "citrus canker" is nothing more than a disguised trade barrier, since the disease has long been eradicated in Mexico.[82]

It can be difficult to distinguish between a legitimate environmental measure and protectionism in disguise. For this reason, both NAFTA and the Uruguay Round of GATT would subject some laws to a science test. NAFTA would require that standards designed to protect human, plant, or animal health, such as pesticide residue limits, be "based on scientific principles" as well as on "risk assessment, as appropriate to the circumstances." The Uruguay Round proposal takes this principle a step further, requiring "scientific justification" as the basis of any such laws deviating from international norms. Because scientists often disagree on what constitutes "justification" (which may be a matter of judgement, even if the scientific findings themselves are not), environmentalists are concerned that the domestic environmental laws of some countries could be subject to challenge.[83]

Even if a policy is genuinely aimed at protecting the environment, it can still have the effect of impeding international trade. A bill requiring that bottles be refillable, for instance, might work to the disadvantage of foreign bottlers who would find it costly to pick up the empties. In such situations, a nettlesome question remains as to which goal—trade or environment—should take precedence.

One of the first prominent tests of this question arose in the late 1980s, when foreign bottlers charged that a Danish law forbidding the selling of beer and soft drinks in cans and requiring that bottles be refillable was a disguised restraint on trade, and thus a violation of EC law. To resolve the matter, the European Commission ultimately turned to the European Court of Justice. In a landmark ruling, the Court found that while the Danish law did disadvantage foreign bottlers, the environmental benefits generally outweighed the constraint on trade. However, the Court did find one relatively minor aspect of the Danish law (requiring bottle types to be officially approved) to be "disproportionate" to the goal at hand.[84]

This ruling implied that under EC law, for an environmental law that constrains trade to be allowed, its environmental benefit must be deemed "proportional" to the restraint on trade—that is, of great enough value to be worth the lost trade. To date, this subjective balancing act has not been part of GATT law or its accompanying code on standards, which has generally required only that a given measure not be a disguised restriction on trade. But some provisions in the Uruguay Round would move GATT in the EC's direction. For instance, in some instances members are expected to choose the least trade-restrictive option available to meet a chosen environmental goal. NAFTA, too, contains some troubling language requiring that certain environmental measures be applied only to the extent "necessary" to achieving a country's chosen level of environmental protection. "Necessity," too, can be subjective, and trade dispute resolution panels have tended to interpret trade agreements in a way that would rule out a number of legitimate environmental policies.[85]

In a landmark ruling, the Court found that the environmental benefits generally outweighed the constraint on trade.

Another environmental concern is that trade agreements generally do not allow product standards imposed on imports to include reference to how something was produced. This became clear in the tuna-dolphin decision of September 1991, when a GATT panel ruled that the U.S. Marine Mammal Protection Act was inconsistent with GATT when it forced an embargo against Mexican tuna. The United States imposed the embargo after determining that the Mexican fishers were catching tuna in a kind of net that kills more dolphins "incidentally" than U.S. fishers are allowed to. The reason for the provision of the U.S. law was that if the United States only restricted tuna caught by its own fishers, then foreign-caught tuna might simply replace the U.S. catch, subverting the purpose of the law. Indeed, there was considerable evidence that before the law was amended to prevent it, many U.S. fishers were simply reflagging their vessels and carrying on as usual, further undermining the goal of saving dolphins.[86]

By one interpretation, the embargo on tuna from Mexico amounted to a product standard—a national law defining conditions for tuna sold in the United States. The dispute panel saw it differently, however. It decreed that the GATT rules, which allow for national product standards to imported goods, did not apply to this case because it was the process by which the tuna was produced rather than the tuna itself that was being rejected by the United States. Though GATT explicitly permits import bans on products made with prison labor, another "process standard," the judges' interpretation recognized no similar provision for environmental standards.[87]

The judges based their ruling on the assumption that the environmental exceptions permitted by GATT only pertained to the environment within domestic borders or jurisdiction. And because the embargo was imposed unilaterally rather than through an international treaty, the panel viewed the U.S. action as an attempt to foist its own environmental laws on another sovereign country. The ruling thus failed to distinguish between environmental issues of purely national concern and those designed to protect the global commons—the oceans and the atmosphere. By applying the "domestic borders" criterion, the judges determined, in effect, that in the absence of an international agreement there is virtually no way to protect the Pacific dolphin, whose habitat is not contained within any country's borders.[88]

For the time being, the Mexicans have decided not to press their case, in all likelihood because they do not want to jeopardize the NAFTA. They chose instead to try to work out a political compromise through which the Marine Mammal Protection Act would be amended so that the embargo could be lifted. So far, however, Mexico has not agreed to a five year moratorium on the killing of dolphins beginning in 1994, as an amendment to the act requires for the embargo to be lifted. Meanwhile, the European Community—the target of a secondary U.S. embargo imposed on countries that purchase the tainted tuna—has now lodged a complaint of its own against the U.S. law. Another dispute panel will hear the case, and the reasoning behind the original tuna-dolphin ruling will be reexamined.[89]

The GATT ruling thus failed to distinguish between issues of purely national concern and those designed to protect the global commons.

If the new panel reinforces the conclusions of the earlier one, several existing or contemplated import bans could be overruled based on the same logic. The threatened laws include, among others, a U.S. ban on imports of shrimp from countries whose governments do not require shrimpers to use the turtle excluder devices required of U.S. shrimpers, and a European Community ban (after 1994) on imports of furs from countries where painful "leg-hold" traps are permitted. The United States has also threatened to embargo fish from countries using drift nets (which have decimated some fish populations) or killing whales, most species of which are in decline. It will be ironic indeed, if animals like dolphins and whales, which live largely in international waters, are abandoned to the demands of commerce on the grounds that their protection violates national sovereignty.[90]

In one of the latest disputes to raise concern, Austria recently attempted to introduce a 70 percent tax on tropical timber, as well as a requirement that tropical timber be labeled as such, with an option to provide a second mark indicating whether or not it came from a sustainably managed stand. The Association of Southeast Asian Nations (ASEAN) complained that the Austrian law was a violation of GATT because it did not apply to temperate timber as well. The ASEAN countries threatened to embargo all Austrian exports, jeopardizing $1.8 billion worth of contracts. Austria has since backed away from the tax proposal, though it is standing by its labeling law.[91]

Another recent skirmish involves European Community complaints that two U.S. automobile taxes intended to promote fuel efficiency—the Corporate Average Fuel Economy Law and the gas-guzzler tax—violate GATT. The EC has said that it plans to call for a GATT dispute resolution panel. The U.S. claims the two taxes would meet GATT tests, as they are applied to domestic producers equally. The E.C. counters that though this appears to be true, they are applied in such a way that European producers are disadvantaged.[92]

Environmentally related export controls, too, can be restricted by trade agreements. When resources are scarce—be they energy, minerals, timber, or water—countries sometimes want to reserve supplies for their own people. This can allow for conservation while still leaving room for domestic consumption. But treating domestic and overseas markets differently runs counter to free trade principles. As a result, such restrictions are discouraged by some trade bodies.

Sometimes, for example, countries restrict exports of unprocessed goods in order to reduce resource depletion while creating jobs and profits in value-added industries. The trouble is that policies to encourage domestic processing can be attacked as trade restrictions. For instance, Canada, Ghana, Indonesia, the Philippines, Thailand, Uganda, and the United States have all banned the export of raw logs under certain circumstances. The European Community charged that the Indonesian ban violated GATT, and perhaps partly as a result, Indonesia recently decided to abandon the policy.[93]

Similarly, until recently, British Columbia had a policy that restricted the export of unprocessed fish in order to enable local fishers, many of them Native Americans, to continue to earn a livelihood while limiting fishing to avoid depleting stocks. A dispute resolution panel convened under the U.S.-Canada free trade agreement declared Canada's policy inconsistent with the agreement. When British Columbia then tried to impose a landing-and-counting requirement that was ostensibly a conservation measure, a GATT panel ruled against the province yet again, arguing that the primary goal of the policy was not protecting fish.[94]

There is no doubt that export restraints clash with free trade notions. In some cases, they may not be the most desirable policies from an environmental perspective either. For example, the wood processing technologies available in developing countries may be older, and therefore more wasteful and polluting, than technologies available elsewhere. And without regulation, the lure of greater profits through processing could cause more resource extraction, not less. Indonesia's ban is believed by many experts to have been environmentally counterproductive for just these reasons. On the other hand, there can be more incentive to protect a natural resource that is yielding a high economic return. And politically, export restraints are often the most effective way to preserve resources, as they can make restrictions on resource extraction possible without undermining the economic base of entire communities.[95]

In the wake of the tuna-dolphin ruling, fear is growing that even international environmental treaties are not immune from challenge by international trade bodies if they contain import or export restrictions. At least 17 treaties fall into this category. In the Basel Convention on hazardous waste export and in the CITES treaty on endangered species, restricting trade is the very *purpose* of the agreement. In other cases, such as the Montreal Protocol on depletion of the ozone layer, restrictions are used to try to prevent countries that have not signed the treaty from undermining its effectiveness—or to enforce compliance by those that have.[96]

All of this presents the world's governments with a momentous legal problem. With GATT aimed at limiting most restrictions on trade, and the environmental treaties requiring them, two sets of international agreements are in head-on conflict. Which should take precedence? Under international law, if both parties to the dispute are members of both agreements, the most recent treaty generally holds sway—which would tend to protect most environmental treaties. But problems could develop if a country not party to the environmental treaty were to argue that GATT should rule—though so far, no such cases have arisen. To thicken the plot still further, successful conclusion to the Uruguay Round might mean that GATT would supplant the environmental treaties as the most recent agreement.[97]

NAFTA, on the other hand, explicitly states that in cases where the obligations of the agreement conflict with the obligations of the Basel, Montreal, and CITES conventions, the environmental treaty shall prevail. Still, if countries have "equally effective and reasonably available" options, they are to use the approach that is "least inconsistent" with NAFTA to meet these treaties' terms. If NAFTA members wish to add any other treaties to this list, all three countries must agree, so conflicts over future treaties could well arise.[98]

One added policy instrument that could be challenged as a violation of free trade rules is the use of subsidies for environmental goals. Trade agreements generally discourage subsidies, and provide for countervailing duties in some cases to compensate for their continued use. On balance, this could benefit the environment if rules are developed to restrict subsidies that cause environmental harm, such as those which promote unsustainable forestry, energy mega-projects, mines, and intensive farming. However, subsidies that are expressly aimed at financing environmental protection could be threatened. Ironically, while NAFTA frowns on *most* subsidies, including environmentally-oriented ones, it winks and looks the other way when the subsidies are for oil and gas exploration. On the other hand, amendments being considered in the Uruguay Round would exempt some agricultural land set-aside programs such as the U.S. Conservation Reserve Program from the threat of countervailing duties. Still, a large number of other more general environmental subsidies would be vulnerable under the latest draft of GATT revisions.[99]

There are a number of steps that could be taken to minimize the potential for environmental trade disputes, and to raise the chances that environmental interests will be adequately weighed when they do arise. The prevailing approach is to attempt wherever possible to harmonize product standards and specifications, which prevents conflicts over different rules. The single market program in Europe includes a far-reaching effort to develop equivalent product standards. It is possible, though not always easy, for countries to impose stricter standards than the EC norm. But in a number of instances, common EC standards have been patterned on the most progressive national laws, in order to minimize barriers to trade while not preventing an individual country from maintaining the high level of protection it has chosen. This leads to an "upward harmonization"—a continent-wide upgrading of environmental standards. That process is under way with EC directives on packaging and on environmental labeling now being developed. By raising every country to the same high standard, it removes the problem of progressive environmental laws being challenged as barriers to trade.[100]

Two sets of international agreements are in head-on conflict.

When harmonization is set at a lower standard (perhaps to accommodate countries with less-developed environmental laws), however, the solution is more difficult. One of the most contentious issues to arise in the debate over the Uruguay Round of GATT concerns the proposal it contains for harmonization of food safety standards, as well as a wide array of other technical standards. Environmental and consumer activists fear this will lead to a "least common denominator" approach, with countries that wish to pursue stricter policies subjected to a variety of legal hoops, such as the scientific justification test. Under NAFTA, a similar effort to harmonize a number of product standards is envisioned. Notably, to avoid the least-common-denominator problem, the text specifically provides for the *upward* harmonization of standards. The NAFTA also contains stronger language than does the GATT on the right of countries and localities to go beyond the harmonized standard if they wish.[101]

Beyond the upward harmonization of standards, trade agreements need to be more explicit in recognizing the need for exceptions to the usual trade rules based on environmental concerns. Under the present terms, the decks are stacked against the environmental cause at the GATT, given the pro-trade bent of the organization and the limited nature of the exceptions to its rules that are granted to preserve human health and natural resources. The word "environment" does not even appear anywhere in the current GATT text. As the tuna-dolphin ruling demonstrated, the agreement glaringly lacks—and needs—a clear stipulation that the process by which production is prepared for export is a legitimate criterion by which to judge whether it should be granted access to one's market. It also lacks—and needs—a clear acknowledgement that certain environmental issues are of global rather than merely domestic concern. The implication of such acknowledgement would be that trade actions to protect these global resources do not amount to unwarranted meddling in the domestic affairs of sovereign nations. Finally, it is critical that the agreement recognize and protect the crucial role trade sanctions can play in enforcing and inducing participation in international environmental agreements.[102]

For cases where inevitable differences do arise, the dispute resolution procedures themselves are of critical importance. A key issue in these cases is where the burden of proof lies. Under GATT, the country whose law is challenged in many cases bears the burden of proving its law is not a disguised form of protectionism, or contrary to the agreement in some other way. The NAFTA will be a modest improvement over GATT in this regard, with the burden of proof carried instead by the challenging country in most cases. Another key issue is the quality of the environmental advice being received by the panels making the decision. Under GATT, environmental expertise has been sorely lacking, with the panels reflecting the strong pro-trade bias of the organization. Under the NAFTA, an expert advisory panel to be created for the dispute resolution proceedings may include environmental advisers, but there is no actual *requirement* that they be included—or that they be consulted by the parties to a dispute. Finally, and most ominously from the standpoint of environmentalists, is the fact that the dispute resolutions are conducted in secrecy. Even the panel decisions are not routinely released to the public, let alone adequate opportunities provided for public comment while the deliberations are underway.[103]

In a sign of its concern over the environmental implications of adverse panel rulings, the European Parliament recently passed a resolution calling for a two year moratorium on all GATT panel judgements affecting the environment, while the pressing issues of GATT reform are confronted.[104]

Political Trading

In a remarkably short time, environmental issues have become a prominent element in world trade negotiations. Both the EC's Maastricht Treaty and the North American Free Trade Agreement recognize the pursuit of sustainable development and strengthened environmental policies to be key goals of the agreements, on a par with expanding trade. This is an achievement that has yet to be matched by GATT. Though these developments are promising, the next step—reforming the agreements themselves to embrace this goal—poses a far greater challenge.[105]

It was not until the late 1980s that the environmental community first turned its attention to GATT. The initial focus was the environmental implications of the Uruguay Round, which at that point had been under negotiation since 1986 and was thought to be nearing completion. Family farm groups voiced concerns that the removal of agricultural subsidies under the agreement would jeopardize their survival. They enlisted environmental support for their campaign, arguing that smaller farms are often more ecologically sustainable than large ones.

Consumer groups, worried that food safety and other domestic environmental laws might be challenged as trade barriers, also became actively involved in the anti-Uruguay Round campaign at about that time. It was the tuna- dolphin ruling of September 1991, however, that provoked the largest cries of alarm, particularly in the United States, and galvanized a movement throughout the international environmental community to address the need for GATT reform. The issue of trade and the environment is now firmly on the international agenda, and a variety of efforts to address it are underway.[106]

In the view of many environmental and consumer groups, the immediate priority is to block adoption of the Uruguay Round unless some fundamental changes are made to the draft of the agreement released by GATT General-Secretary Arthur Dunkel in December 1991, that is now serving as the basis of negotiations. Key concerns include the provisions on standards and the plans to create a new Multilateral Trade Organization that would have more binding powers than existing dispute settlement panels to overturn domestic environmental laws. In response to these concerns, U.S. negotiators have made efforts to reopen the negotiations in these areas. However, other countries may resist reopening these portions of the text at this late stage.[107]

Over the longer term, it is now widely acknowledged that some environmental reform of the GATT is needed. Though environmental groups would like to see this happen before expanding the agreement further through the Uruguay Round, most governments feel that the priority is to complete the current round before embarking on any "green round".[108]

In one effort that may lay the groundwork for a future green round, a GATT working group on trade and the environment—a group that had been created in the early 1970s but had never met—was reactivated in November 1991. Its agenda includes the relationship between GATT rules and the trade provisions of international environmental agreements, the notification of national environmental regulations with trade effects, and the trade effects of packaging and labelling requirements that are proliferating around the world. So far, the talks have not yielded any results.[109]

Governments have also grappled with the question of trade and the environment in other forums. It was a topic of exhaustive discussion at the Earth Summit in Rio de Janeiro in June 1992. In most of the documents that emerged from the conference, governments affirmed the principle that freer trade and environmental protection are mutually supportive goals. Unfortunately, the delegates took this as an article of faith and did little to specify just how to make this principle govern real-world commercial activity. Charles Arden-Clarke of the Worldwide Fund for Nature ruefully observed: "The whole of the Agenda 21 section dealing with trade amounted to an evasion of key trade and environment issues, rather than a basis for their solution." Perhaps most alarmingly, in a rebuke to the U.S. position on the tuna-dolphin controversy, the Earth Summit documents hold that the use of unilateral trade measures is to be avoided. In follow-up discussions at the GATT, the United States has continued to proclaim the right to take unilateral actions—to strong opposition from much of the rest of the world.[110]

Ongoing discussions among the members of the OECD on trade and the environment may lay the groundwork for an eventual "greening of the GATT." The starting point for the talks are the 1972 OECD "Guiding Principles Concerning the International Economic Aspects of Environmental Policies," which were the last attempt to lay down some rules. In the 1972 guidelines, OECD members committed themselves—in theory, at least—to the "polluter pays" principle, which encourages cost internalization as a means of avoiding trade distortions. The Earth Summit's Rio Declaration affirmed the polluter pays principle at the international level. But the polluter does not yet pay full environmental costs, as is amply demonstrated by the high levels of air and water pollution and waste generation around the world, and by the continued erosion of the earth's natural resource base.[111]

The 1972 guidelines also endorse harmonization of standards where possible, while noting that countries should have the right to choose the level of environmental protection they deem desirable. The guidelines discourage the use of border tariffs to compensate for cases where trading partners are deriving competitive advantage through weaker environmental laws.[112]

In the current updating, the OECD governments are aiming to negotiate two documents—a set of guidelines on how to minimize the negative trade effects of environmental policies, and one on how to avoid negative environmental effects of trade policies. The intention was to have the guidelines completed by spring of 1992, and then by the spring of 1993. However, progress has been slow. Unlike in the 1972 discussions, there has been an explicit effort to look at the issue from an environmental as well as a trade perspective. However, agreement appears closest on measures that would insure that environmental measures do not impede trade, rather than on those that would try to insure that trade does not harm the environment.[113]

One issue that governments have yet to squarely confront at the international level is the question of what the proper institutional forum is for considering trade and environment questions. Certainly GATT will play a strong role. If the envisaged Multilateral Trade Organization is ever created, it will be critical that it include a strong commitment to the tenets of sustainable development. But many question whether or not an organization as dedicated to free trade as the proposed organization—or GATT—could ever approach environmental issues from a non-biased perspective. For this reason, it may be appropriate to turn to the newly created U.N. Commission on Sustainable Development for environmental oversight of trade bodies. Alternatively, perhaps a strengthened U.N. environment agency could be dedicated to this purpose.[114]

As the talk continues at the international level, there is considerable movement at the regional level to develop policies that integrate trade and the environment. In the North American Free Trade Agreement and the European Maastricht Treaty, particularly, the precedents being set now are likely to become models for other regional trade pacts in the future.[115]

The fate of the NAFTA is somewhat in limbo, given the presidential transition in the United States. The negotiations were completed in the fall of 1992, in the midst of the election campaign, and the agreement was signed December 17 by the outgoing President Bush, Mexican President Carlos Salinas de Gortari, and Canadian President Brian Mulroney. President Clinton has said he will not reopen the text, but has voiced support for the negotiation of a side agreement on the environment to ensure that environmental laws in all three countries are adequately enforced. He has also committed to creation of a North American Commission on the Environment, with the authority to impose penalties for violations of the NAFTA's environmental terms.[116]

A key question is how much power will be vested in the new environmental commission to be created under NAFTA. Most environmental groups envision the Commission monitoring the enforcement of environmental laws, perhaps including provisions for inspections. Some are pushing for the Commission to play an active role in promoting the upward harmonization of environmental laws from the Arctic to the Yucatan. Yet to be determined is whether the new Commission will have the power to levy fines or impose other types of penalties for violations of the agreement's environmental provisions. However, the Mexican government is concerned that a body with these kinds of powers would infringe on its sovereignty. Whether or not the environmental community will throw its considerable political weight behind the agreement will largely depend on the content of these side agreements.[117]

In the European Community, the number of environmental noncompliance complaints rose from 11 in 1984 to 480 in 1990.

Another major concern is the need for adequate funding to help Mexico meet the costs of its stepped-up environmental obligations. The Institute for International Economics suggests that a fund of at least $5 billion will be required to mitigate the existing environmental problems and prevent future ones in the border area alone. Other estimates are substantially higher. To raise the needed money, some analysts have suggested that a portion of tariff earnings from cross- border transactions go into a fund for environmental and social adjustment that would be phased out gradually as tariffs are eliminated. The Mexican government is considering levying a "user fee" on both imports and exports to raise money for environmental enforcement and cleanup. Another proposal calls for creating a development bank for the region modeled on the European Bank for Reconstruction and Development. Of course, funding can be generated via more traditional methods as well. The U.S. and Mexican governments have already agreed to a border cleanup plan of at least $850 million over four years that will fund a number of pollution control initiatives and cooperative programs in the region. Much of the money has not yet been appropriated, however.[118]

One obvious model for the NAFTA is the European Community, which has a long history of working to reconcile trade interests with environmental and labor concerns. Created in the 1950s as a free trade agreement in coal and steel that its founder, Jean Monnet, hoped would harness economic interests to keep the peace between Germany and France, the EC has of course developed into far more than a mere free trade agreement. From early on, its leaders agreed that economic integration in Europe could not proceed without social and political integration, or the result would be a race to the bottom in labor and environmental standards. They also agreed that creation of a true community would require addressing the large disparities in wealth between the northern and southern parts of the continent.[119]

The EC's institutional machinery for environmental protection serves as a useful starting point for envisioning what the role of the North American Commission on the Environment will be. The various branches of the European Community, including the democratically elected European Parliament, all play a role in the promulgation of hundreds of common minimum environmental standards on products and production. EC "directives" are enforceable at the European Court of Justice.

Nongovernmental groups play an active role in the process: they can and do submit charges of noncompliance. The number of such complaints rose from 11 in 1984 to 480 in 1990. Though the European Court has little fining or penalty power, the public shaming that results from a ruling against a country can be enough to persuade transgressors to change their ways. And in some limited cases, EC laws may be enforceable in national courts. Financing for environmental protection in the Community's less-affluent members (such as Spain and Portugal) comes through two main international channels: a multi-billion dollar "adjustment fund" that includes some environmental projects, and a $552-million fund (to be dispersed over three years) dedicated specifically to financing the development and implementation of EC environmental policy.[120]

The Maastricht treaty will strengthen the Community's hand in the environmental arena in a number of ways. Among the most important innovations will be inclusion of the cause of sustainable development in the EC's Treaty of Rome, more provision for agreeing to environmental laws by majority rather than unanimous voting, a greater voice for the European Parliament in making EC environmental laws, and limited power to impose financial penalties in cases where countries fail to comply with rulings of the European Court of Justice.[121]

Reconciling Trade and the Environment

From antiquity to the present, trade has played a central role in human development—shaping such historical trends as the spread of Islam into Africa, the rise of European city-states, the European expansion into the Americas, and the emergence of Japan as a new kind of global superpower. Human cultures have always accorded great importance to the business of exchanging goods with one-another, based on their varying natural and acquired endowments. That importance is only likely to increase.[122]

Just as trading patterns have contributed to historic sea changes in the patterns of human habitation and industry, they are now becoming a driving force in one of the latest challenges to confront the human species, and arguably the greatest: the need to reverse the deterioration of the natural resource base upon which all economic activity—and life itself—ultimately depends.

Though trade is as old as humankind itself, governments have pursued the goal of expanding international commerce with particular vigor in the half-century since the end of World War II. The result has been a dramatic acceleration of the millennia-long evolution of trade—from a mode of exchange between individuals, to one that linked communities, then city-states, and so forth—to the current global marketplace in which goods of all sorts are shipped thousands of kilometers, often at jet-speed, and where capital is instantaneously mobilized at the punch of a computer keyboard.

Most economists laud this rapid expansion as one of the main engines of the dazzling economic growth that has occurred around the world during this period. But the Gross National Product, the conventional measure of economic growth, is flawed, as it does not account for environmental depreciation or reflect distribution of wealth. World Resources Institute economist Robert Repetto has found, for instance, that the average annual growth of Indonesia's GNP between 1971 and 1984 was only 4 percent instead of the official 7.1, when the deterioration of its forest, soil, and petroleum assets are figured into the equation. Similarly in the Philippines, resource depreciation in just three sectors—forests, soils, and coastal fisheries—averaged more than 4 percent of GDP from 1970 to 1987, and 20 percent of gross investment. And though economic growth in recent decades has enriched many, there are hundreds of millions for whom living conditions have deteriorated rather than improved. To the extent that world trade has contributed to ecological deficit-financing, or to worsening income inequality, it has failed to promote a secure economic future.[123]

Probably the single most important reform needed to place world trade on a more sustainable footing is widespread internalization of environmental costs through tax policy and other measures. By increasing the price of coal-based electricity, for instance, a tax on carbon dioxide emissions might give photovoltaic cell-derived solar electricity just the edge needed to be competitive either domestically or internationally, helping to stave off the threat of global warming. Similarly, if wood from primary forests were priced to reflect the loss of biological diversity suffered when it is harvested, timber from sustainably managed stands would be more competitive in national and global markets.

Unfortunately, encouraging more widespread internalization of environmental costs even at the national level is proving to be a mammoth political challenge, as demonstrated by the battle over energy taxes in the United States. If developing countries are to internalize the costs of environmental damage into the prices of their products, when industrial countries have been effectively passing these costs off to future generations for decades, they will need considerable financial and technical assistance from richer nations to assist in the transition.

Yet to adequately confront the environmental challenges posed by world trade, costs must be internalized not only within nations, but between them as well. Otherwise, consumers in one country will purchase imported products made artificially cheap by the failure of another country to adequately account for their real costs. Countries making a serious effort to move toward prices that reflect full environmental liabilities might find themselves at a chronic disadvantage in international markets. The prospect of that could discourage them from adopting the policy in the first place.

The demands of world trade are thus likely to spur movement toward internationally-agreed measures to internalize costs—be they taxes, regulations, or labeling programs. In effect, this means political integration must follow economic union. This parallels developments in the United States, where federal worker and environmental legislation is justified by the Interstate Commerce Clause in the U.S. Constitution, on the grounds that it is necessary to leveling the playing field for trade between States. As President Franklin Roosevelt told the U.S. Congress in 1937 when promoting the Fair Labor Standards Act that created the minimum wage and child labor laws, "Goods produced under conditions which do not meet a rudimentary standard of decency should be regarded as contraband and ought not be allowed to pollute the channels of interstate trade." It also parallels developments in the European Community, and could well foreshadow integration promoted by the North American Free Trade Agreement. Political integration may develop in other regions as well—including the Asian trade bloc now in the early stages of negotiation.[124]

Though environmental integration will be pioneered at the regional level, where trade pacts are proliferating and countries have more cultural or political ties to begin with, it will ultimately be experienced globally as well. This will happen both through the further development of a body of international environmental law (through international institutions such as the U.N. Commission on Sustainable Development), and through the greening of the GATT and other trade bodies.

The prospects that trade considerations could drive a ratcheting-up of environmental laws around the world offer some momentous opportunities. But any optimism that this will happen easily can be dashed by a single observation: both the rules of trade and the institutions that implement them were largely created with the narrowly-defined interests of global business in mind, and they have scant—if any—provisions for public participation. In stark contrast to many international environmental negotiations, trade negotiations and dispute resolution proceedings are conducted in complete secrecy, with all documents unavailable to the public except through leaks to the press. Business groups have traditionally had far greater access to the trade negotiating arena than have environmental and other public interest groups.

Private citizens themselves have shied away from becoming adequately informed about the work of these organizations, as the bureaucratic language and obfuscating nature of the proceedings tend to obscure their importance. People have simply not known that such fundamental matters as the wholesomeness of the food they eat, the integrity of the communities they live in, and, indeed, the health of the planet itself are at stake. If trade is to become ecologically and socially sustainable, its isolation from public scrutiny will have to end. "Either the progressive systems of the advanced industrial democracies will spread into the developing world or the Third World will move north," wrote Walter Russell Mead in *Harper's* last year. "Either Mexican wages will move up or American wages will move down. Environmentalists, labor unions, consumer groups, and human-rights groups must go global—just as the corporations have done." These organizations have in fact begun this process. Meetings between Canadian, Mexican, and U.S. citizen groups now occur on a regular basis to plot a joint strategy toward the NAFTA.[125]

Despite the warnings of scientists that it is becoming increasingly urgent for the world to shift to a sustainable economy—and to a sustainable basis for the international trade that increasingly drives it—some entrenched industrial and political interests remain unconvinced. Some even remain complacent, confident that the promise of "new technology" or "human ingenuity" will solve even the most daunting of problems in due course.

But for those who harbor any doubt that the cost of business will have to be redefined comprehensively and soon, it is well worth taking an unblinking look at the alternative. What would be the cost of *not* eliminating the incentives that now encourage the liquidation of global assets for the short-term profits of a few nations or corporations? With global trade now expanding at a rate that dwarfs most other measures of growth—faster than global GNP, and far faster than the human population itself—the consequences of international "business as usual" are frightening indeed.[126]

Trade, of course, is neither inherently good nor bad; but how it is conducted in the future is now a matter of deep concern—and unprecedented opportunity. Trade can spread wealth and knowledge—or it can accelerate the destructive trends now under way. To those who have come to recognize that the list of living systems and materials that make up the Earth's environment constitute the global economy's main inventory of economic assets, there is no question what the choice must be.

Notes

1. World GNP growth from Hal Kane, "Economy Contracts Slightly," in Lester R. Brown, Christopher Flavin, and Hal Kane, Worldwatch Institute, *Vital Signs 1992: The Trends That Are Shaping Our Future* (New York: W.W. Norton & Co., 1992); Information on services from General Agreement on Tariffs and Trade (GATT), *International Trade 90-91, Vol. II* (Geneva: 1992), and from United Nations Development Programme (UNDP), *Human Development Report 1992* (New York: Oxford University Press, 1992); Investment growth from United Nations Center on Transnational Corporations, *World Investment Report 1991: The Triad in Foreign Direct Investment* (New York: 1991).

2. Nicholas Lenssen, "Third World PVs Hit the Roof," *World Watch*, May/June 1992.

3. GATT, *GATT: What It Is, What It Does* (Geneva: 1991).

4. David Marsh and Andrew Hill, "Europe 1993: The Single Market and You," *Financial Times*, January 4, 1993; Barbara Verhoeve, Graham Bennett, and David Wilkinson, *Maastricht and the Environment* (Arnhem and London: Institute for European Environmental Policy, 1992).

5. For a basic description of the NAFTA, see Gary Clyde Hufbauer and Jeffrey J. Schott, *NAFTA: An Assessment* (Washington, D.C.: Institute for International Economics (IIE), 1993); "Canada, Mexico Stress Starting Nafta on Time," *Journal of Commerce*, February 17, 1993.

6. United Nations Conference on Environment and Development (UNCED), Rio de Janeiro, June 3-14, 1992, *Agenda 21* (New York: United Nations Publications: 1992).

7. For statements of the need for trade to contribute to the goal of sustainable development, see Ibid.; Organisation for Economic Co-operation and Development (OECD), "Trade and Environment: A Progress Report," Paris, May 1992; and United Nations Conference on Trade and Development, "The Outcome of UNCTAD VIII in the Field of Sustainable Development and the Environment," United Nations, undated.

8. GATT, op. cit., note 1. Table 2 based on U.N. Food and Agriculture Organization (FAO), *Forest Products Yearbook 1990* (Rome: 1992); FAO, *Production Yearbook 1990* (Rome 1991); FAO *Trade Yearbook 1990* (Rome: 1991); FAO, *Fishery Statistics Yearbook 1989* (Rome: 1991); Fearnresearch, *World Bulk Trades 1990* (Oslo: 1991); and United Nations, *1990 Energy Statistics Yearbook* (New York: 1992). Table 3 based on the following: Commodity export earnings calculated from International Monetary Fund (IMF), *International Financial Statistics Yearbook 1991* (Washington, D.C.: 1991) and FAO, *Trade Yearbook 1990* (Rome: 1991). Otherwise, gold and diamonds from Michael Heydari, U.S. Bureau of Mines, private communication, February 8, 1993; coffee, banana, cotton, tobacco, and beef from FAO, *Production Yearbook 1991* (Rome, 1992) and from FAO, *Trade Yearbook 1990* (Rome: 1991); sugar from USDA, Economic Research Service, *Sugar and Sweetner: Situation and Outlook Report* (Washington, D.C.: 1990); petroleum from U.S. Department of Energy (DOE), *International Energy Annual 1990* (Washington,

D.C., 1992); fish from Arnór Halldórsson, Icelandic Ministry of Fisheries, private communication February 11, 1993; and logs and timber from FAO, *Forest Products Yearbook 1990* (Rome: 1992).

9. GATT, op. cit., note 1; Jim MacNeill et al., *Beyond Interdependence: The Meshing of the Earth's Economy With the Earth's Ecology* (New York: Oxford University Press, 1991).

10. FAO, *Trade in Tropical Timber: Monthly Bulletin: Annual Totals* (Rome: various years); MacNeill et al., op. cit., note 9; Sandra Postel and John C. Ryan, "Reforming Forestry," in Lester R. Brown et al., *State of the World 1991* (New York: W.W. Norton & Co., 1991); FAO, *Forest Products Yearbook 1990*, op. cit., note 8.

11. Postel and Ryan, op. cit., note 10.

12. FAO, *Trade in Tropical Timber: Monthly Bulletin: Annual Totals* (Rome: 1988).

13. FAO, *Forest Products Yearbook 1990*, op. cit., note 8; John C. Ryan, "Plywood vs. People in Sarawak," *World Watch*, January/February 1991; Victor Mallet, "Guarding a Dwindling Asset," *Financial Times*, August 28, 1992.

14. Meri McCoy-Thompson, "Sliding Slopes Break Thai Logjam," *World Watch*, September/October 1989; Victor Mallet, "Cambodia Powerless to Prevent Plundering of Its Forests," *Financial Times*, August 20, 1992; William Branigan, "Phnom Penh Said to Undercut U.N. Effort to Save Forests," *The Washington Post*, February 3, 1993.

15. GATT, op. cit., note 1.

16. Michael Zielenziger, "Free-Trade Agreement Could Uproot Thousands of Mexican Farm Workers," *Journal of Commerce*, August 18, 1992; David Clark Scott, "Mexico's Reforms Transform Agrarian Sector," *Christian Science Monitor*, November 12, 1992; Mark Ritchie, "Free Trade versus Sustainable Agriculture: The Implications of NAFTA," *The Ecologist*, September/October 1992; Charles Arden-Clarke, World Wide Fund for Nature (WWF)-International, Gland, Switzerland, "Agriculture and Environment in the GATT: Integration or Collision," undated, published in *ECOS*, Vol. 13, No. 3, 1993.

17. "Trade and the Environment," in GATT, *International Trade 90-91, Vol. I* (Geneva: 1992); U.S. Congress, Office of Technology Assessment (OTA), *Trade and the Environment: Conflicts and Opportunities* (Washington, D.C.: U.S. Government Printing Office, 1992); Robert Repetto, World Resources Institute, memo to the NACEPT Trade and Environment Committee, December 16, 1992.

18. Annual sales loss from UNDP, *Human Development Report 1991* (New York: Oxford University Press, 1991); Charles Arden-Clarke, *The General Agreement on Tariffs and Trade, Environmental Protection and Sustainable Development* (Gland, Switzerland: WWF-International, 1991); Jodi L. Jacobson, *Gender Bias: Roadblock to Sustainable Development*, Worldwatch Paper 110 (Washington, D.C.: Worldwatch Institute, September 1992).

19. GATT, op. cit., note 17; OTA, op. cit., note 17; Repetto, op. cit., note 17; Amy Kaslow, "Farm Subsidies Create Bounty, Boondoggles," *Christian Science Monitor*, November 4, 1992.

20. Letter of Justin Ward, Senior Resource Specialist, Natural Resources Defense Council, Washington, D.C., to Carla A. Hills, U.S. Trade Representative, and Edward Madigan, U.S. Secretary of Agriculture, March 20, 1992; Mark Ritchie, "GATT, Agriculture, and the Environment: The US Double Zero Plan," *The Ecologist*, November/December 1990.

21. Alan B. Durning, "Fat of the Land," *World Watch*, May/June 1991; Alan B. Durning and Holly B. Brough, *Taking Stock: Animal Farming and the Environment*, Worldwatch Paper 103 (Washington, D.C.: Worldwatch Institute, July 1991); FAO, *Production Yearbook 1990*, op. cit., note 8; FAO, *Trade Yearbook 1990*, op. cit., note 8.

22. Bruce Byers, "Roots of Somalia's Crisis," *Christian Science Monitor*, December 24, 1992; International Monetary Fund, *Balance of Payments Statistics Yearbook*, Vol. 42, Part 1 (Washington, D.C.: 1991).

23. Hal Kane, "Fishing For Trouble," *World Watch*, November/December 1992; FAO, *Fisheries Statistics 1989*, op. cit., note 8; "Iceland Faces Calls for 40% Cut In Cod Catch," *Financial Times*, June 3, 1992; Robert Corzine, "Iceland Finds Enemy to Replace Soviet N-subs," *Financial Times*, October 8, 1992.

24. Figure on developing countries from Nicholas Lenssen, "The Ocean Blues," *World Watch*, July/August 1989; Kelly Haggart, "The Global Scramble for Fish," and "Exporting More, Eating Less," *Panoscope*, November 1992.

25. Chris Hellier, "The Mangrove Wastelands," *The Ecologist*, March/April 1988; Robin Broad and John Cavanagh, *Plundering Paradise: The Struggle for the Environment in the Philippines* (Berkeley: University of California Press, 1993).

26 .John E. Young, *Mining the Earth*, Worldwatch Paper 109 (Washington, D.C.: Worldwatch Institute, July 1992).

27. Ibid.; Information Research Limited, "Trends in Relocation of Potentially Polluting Chemical Industries" (draft), London, June 1992.

28. United Nations, *1990 Energy Statistics Yearbook*, op. cit., note 8.

29. Lucille Langlois, "The Environmental Effects of Trade in the Energy Sector," OECD, Paris, forthcoming.

30. Sally Johnson, "Indians in Canada Cross Border to Fight Utility," *New York Times*, June 4, 1989; British Petroleum, *BP Statistical Review of World Energy* (London: various years); Murray Feshbach and Alfred Friendly, Jr., *Ecocide in the USSR: Health and Nature Under Siege* (New York: Basic Books, 1992).

31. F. Schmidt-Bleek and H. Wohlmeyer, *Trade and the Environment: Report on a Study* (Laxenburg, Austria: International Institute for Applied Systems Analysis and the Austrian Association for Agricultural Research, 1991).

32. Charles Arden-Clarke, *South-North Terms of Trade, Environmental Protection and Sustainable Development*, (Gland, Switzerland: WWF-International, February 1992); "Synthesis Report: The Environmental Effects of Trade," OECD, Paris, forthcoming; Young, op. cit., note 26; Charles Arden-Clarke, WWF-International, private communication, February 17, 1993.

33. GATT, op. cit., note 1.

34. John C. Ryan, "Goods From the Woods," *World Watch*, July/August 1991.

35. Ibid.; Matthew Kaminski, "A Hard Nut to Crack," *Financial Times*, December 2, 1992.

36. Vicki Elkin, "Fairer Trade," *World Watch*, July/August 1992; "Drive to Give Third World Farmers Fair Deal," *Financial Times*, September 16, 1992; Charles W. Thurston, "Socially Conscious Importers Promote Third World Goods," *Journal of Commerce*, August 6, 1992.

37. Gunter Pauli, President, Ecover, Malle, Belgium, private communication, February 11, 1993.

38. Postel and Ryan, op. cit., note 10; Bronwen Maddox, "Catch-all Forestry Rules Urged," *Financial Times*, January 27, 1993.

39. Adam Raphael, "Earth Summit Casts Shadow Over Timber Talks," *Financial Times*, May 15, 1992; "Little Progress Made at Meeting on Managed Logging, Trade in Tropical Timber," *International Environment Reporter*, December 2, 1992; "Debate Said Ahead on Expanding Tropical Timber Pact to Temperate Regions," *International Environment Reporter*, October 7, 1992; Maddox, op. cit., note 38; United Nations Conference on Trade and Development, "International Agreement on Jute and Jute Products" (1989) and "International Sugar Agreement" (1992).

40. OTA, op. cit., note 17; Arden-Clarke, op. cit., note 32; Konrad Von Moltke, "Free Trade and Mutual Tariffs: A Practical Approach to Sustainable Development," *EcoDecision*, June 1992. Export tariffs are consistent with GATT in its current form, while import tariffs are not.

41. Nicholas F. Brady, "The United States and Latin America: A Mutual Partnership for Progress," Speech at the Inter-American Development Bank, Washington, D.C., January 11, 1993, published in U.S. Department of the Treasury Press Release, Washington, D.C., January 11, 1993; Wilfrido Cruz and Robert Repetto, *The Environmental Effects of Stabilization and Structural Adjustment Programs: The Philippines Case* (Washington, D.C.: World Resources Institute, 1992); David Reed, Ed., WWF-International, *Structural Adjustment and the Environment* (Boulder, Colo.: Westview Press, 1992).

42. Jim Vallette and Heather Spaulding, *The International Trade in Wastes: A Greenpeace Inventory* (Washington, D.C.: Greenpeace, 1990); for numerous ongoing instances of waste export see Greenpeace's *Toxic Trade Update*, Second Quarter 1992.

43. United Nations Environment Programme, "Basel Convention on the Control of Transboundary Movements of Hazardous Wastes and their Disposal," Final Act, Basel, Switzerland, March 21, 1989; "United Nations Officials See Basel Treaty As `Limping' Into Effect With Limited Support," Special Report, *International Environment Reporter*, May 6, 1992; "Basel Convention Now in Force," *Toxic Trade Update* (Greenpeace USA), Second Quarter 1992; "United States Blocks Efforts to Prohibit Global Waste Dumping by Industrial States," Greenpeace Press Release, Piriapolis, Uruguay, December 2, 1992; "Conference on Toxic Waste Ends," *Wall Street Journal*, December 7, 1992; "Basel Convention Parties End Meeting Without Call for Total Ban on Toxics Trade," *International Environment Reporter*, December 16, 1992.

44. Task Force on Environment and the Internal Market, *"1992": The Environmental Dimension* (Bonn: Economica Verlag, 1990); Simon Carroll, *The Single European Dump* (Amsterdam: Greenpeace International, 1991); Office of the U.S. Trade Representative (USTR), "Review of U.S.-Mexico Environmental Issues," Washington, D.C., February 25, 1992.

45. "EC Court Ruling May Allow Restrictions on Waste Imports," *International Environment Reporter*, July 15, 1992; "Euro-Court Rules on Waste Trade" and "EC Waste Trade Legislation Passes to U.K. as France Stands Against Waste Trade," *Toxic Trade Update* (Greenpeace USA), Second Quarter 1992; David Gardner, "EC Deal on Toxic Waste Disposal," *Financial Times*, October 21, 1992.

46. Jonathan Dahl, "Canada Encourages Mining of Asbestos, Sells to Third World," *Wall Street Journal*, September 12, 1989; Mario Epelman, "The Export of Hazard in the Third World: The Case of Asbestos in Latin America," *New Solutions*, Summer 1992; Ray Sentes, Professor of Political Science, University of Regina, Saskatchewan, Canada, private communication, October 15, 1992; Paul Rauber, "New Life for White Death," *Sierra*, September/October 1991.

47. U.S. General Accounting Office (GAO), "Export of Unregistered Pesticides is Not Adequately Monitored by EPA," Washington, D.C., April 1989; "'Silent Spring' in the Third World," *Toxic Trade Update* (Greenpeace USA), Second Quarter 1992.

48. "Network Provides Importing Countries with Means to Prohibit Banned Chemicals," *International Environment Reporter*, December 16, 1992; "GATT Argues Over Response to Earth Summit on Trade and Environment," *Inside U.S. Trade*, October 30, 1992; "U.S. Holding Up Agreement on Hazardous Substances," *Third World Economics*, July 1-15, 1991.

49. Ann Misch, "Can Wildlife Traffic Be Stopped?" *World Watch*, September/ October 1992; Sarah Fitzgerald, *The International Wildlife Trade: Whose Business Is It?* (Washington: World Wildlife Fund, 1989).

50. "World Trade in Wildlife," Fact Sheet, TRAFFIC USA, World Wildlife Fund, Washington, D.C., February 1992; Fitzgerald, op. cit., note 49; Misch, op. cit., note 49.

51. Task Force, op. cit., note 44; Debra A. Rose, "A North American Free Trade Agreement: The Impacts on Wildlife," World Wildlife Fund, Washington, D.C., 1991.

52. OECD, *The OECD Environment Industry: Situation, Prospects and Government Policies* (Paris: 1992); capacity of wind turbine exports from Strategies Unlimited, *International Market Evaluations: Wind Energy Prospects* (Mountain View, Calif.: California Energy Commission, undated); value of wind turbine exports from Kevin Rackstraw, American Wind Energy Association, Washington, D.C., private communication, August 21, 1992, based on ibid.; photovoltaic exports are a Worldwatch Institute estimate, based on Rick Sellers, Solar Energy Industries Association, Washington, D.C., private communication, September 9, 1992.

53. Amy Vickers, Amy Vickers & Associates, Boston, Mass., private communication, October 16, 1992; Curtis Moore, "Bush's Nonsense on Jobs and the Environment," *New York Times*, September 29, 1992.

54. Anya Schoolman, U.S. Environmental Protection Agency, private communication, February 18, 1993; "Proposal For a National Environmental Technology and Business Competitiveness Act of 1993," Office of U.S. Congressman Gerry Studds.

55. Robert Thompson and David Lascelles, "Tax May Run Into Political Sands," *Financial Times*, May 15, 1992; Bruce Barnard, "EC Executive Seeks to Impose Energy Tax to Curb Emissions," *Journal of Commerce*, May 13, 1992.

56. United Nations Industrial Development Organization (UNIDO), *Industry and Development: Global Report 1990/91* (Vienna: 1990); UNDP, op. cit., note 18. Two recent studies examine the question of location trends in "environmentally dirty" industry. One paper shows the importance of "environmentally dirty goods" in production in various countries, demonstrating that exports are declining somewhat in the industrial world and growing in the developed. A second notes a growth in toxic-intensive industries in developing countries, particularly ones rather protected from the international market. Patrick Low and Alexander Yeat, "Do `Dirty' Industries Migrate," and Robert Lucas et al., "Economic Development, Environmental Regulation and the International Migration of Toxic Industrial Pollution: 1960-1988," both in Patrick Low, ed., *International Trade and the Environment*, World Bank Discussion Papers (Washington, D.C.: World Bank, 1992).

57. H. Jeffrey Leonard, *Pollution and the Struggle for the World Product* (Cambridge: Cambridge University Press, 1988); OTA, op. cit., note 17; U.S. Bureau of the Census, *Manufacturer's Pollution Abatement Capital Expenditures and Operating Costs* (Washington, D.C.: 1988), cited in USTR, op. cit., note 44; Schmidt-Bleek and Wohlmeyer, op. cit., note 31.

58. Leonard, op. cit., note 57.

59. Roberto Sanchez, "Health and Environmental Risks of the Maquiladora in Mexico," *Natural Resources Journal*, Winter 1990; GAO, "Some U.S. Wood Furniture Firms Relocated from Los Angeles Area to Mexico," Washington, D.C., April 1991.

60. Gary Clyde Hufbacher and Jeffrey J. Schott, *North American Free Trade: Issues and Recommendations* (Washington, D.C.: Institute for International Economics, 1992).

61. Justin Ward and Glenn T. Prickett, "Prospects for a Green Trade Agreement," *Environment*, May 1992; "Nafta May Bring 'Import' of Pollution to Mexico," *Journal of Commerce*, August 20, 1992; "Hazardous Waste from U.S.- Owned Plants in Mexico Dumped Illegally, Panel Told," *International Environment Reporter*, December 4, 1991; GAO, "U.S.-Mexico Trade: Assessment of Mexico's Environmental Controls for New Companies," Washington, D.C., August 1992.

62. S.J. Lewis et al., *Border Trouble: Rivers in Peril* (Boston: National Toxics Campaign Fund, 1991); Council on Scientific Affairs, American Medical Association, "A Permanent U.S.-Mexico Border Environmental Health

Commission," Chicago, June 27, 1990; John Holusha, "Trade Pact May Intensify Problems at the Border," *New York Times*, August 20, 1992; USTR, op. cit., note 44.

63. "Japan, Malaysia Embroiled in Dispute Over Alleged Pollution From Chemical Firm," *International Environment Reporter*, July 29, 1992; Mohammed Motala, "Bayer Poisons South Africa" and "Thor Chemicals Reopens Mercury 'Recycling' Facility in South Africa," *Toxic Trade Update* (Greenpeace USA), Second Quarter 1992.

64. Steve Charnovitz, "Environmental and Labour Standards in Trade," *World Economy*, May 1992.

65. Terry McGowan, Manager of Applications Development, General Electric Lighting, Cleveland, Ohio, private communication October 14, 1992; David Wheeler and Paul Martin, "Prices, Policies, and International Diffusion of Clean Technology: The Case of Wood Pulp Production," and Nancy Birdsall and David Wheeler, "Trade Policy and Industrial Pollution in Latin America: Where Are the Pollution Havens?" in Low, op. cit., note 56.

66. Birdsall and Wheeler, op. cit., note 65.

67. Patricio Chirinos, Secretary of Urban Development and Ecology, speech on Mexican Integrated Environmental Border Plan, Ciudad Juarez, Mexico, October 23, 1991; Governments of Canada, Mexico, and the United States of America, "The North American Free Trade Agreement," September 6, 1992; Sierra Club, "Environmental Concerns Regarding the December 17, 1992 North American Free Trade Agreement," mimeograph, Washington, D.C., undated; Bill Snape, Defenders of Wildlife, private communication, February 14, 1993.

68. On the ILO, James Avery Joyce, *World Labor Rights and Their Protection* (London: Croom Helm, 1980); Charnovitz, op. cit., note 64; George E. Brown, Jr.; J. William Goold, and John Cavanagh, "Making Trade Fair," *World Policy Journal*, Spring 1992; David Clark Scott, "Under Pressure, Guatemala Addresses Labor Law Abuses," *Christian Science Monitor*, December 15, 1992; N. Vasuki Rao, "India and Bangladesh Act to Curb Use of Child Labor," *Journal of Commerce*, January 21, 1993.

69. Hilary F. French, The E.C.-Environmental Proving Ground," *World Watch*, November/December 1991; Cameron Keyes, *The European Community and Environmental Policy: An Introduction for Americans* (Baltimore, Md.: WWF, 1991); Philippe Sands, "European Community Environmental Law: The Evolution of a Regional Regime of International Environmental Protection," *Yale Law Journal*, Vol. 100, No. 1, 1991; Stewart J. Hudson and Rodrigo J. Prudencio, National Wildlife Federation, Washington, D.C., "The North American Commission on Environment and Other Supplemental Environmental Agreements: Part Two of the NAFTA Package," February 4, 1993; Hufbauer and Schott, op. cit., note 5; Hilary F. French, *After the Earth Summit: The Future of Environmental Governance*, Worldwatch Paper 107 (Washington, D.C.: Worldwatch Institute, March 1992).

70. Walter and Loudon study cited in World Commission on Environment and Development, *Our Common Future* (New York: Oxford University Press, 1987).

71. OTA, op. cit., note 17; Arden-Clarke, op. cit., note 18; Steve Charnovitz, "The Influence of International Labour Standards on the World Trading Regime: A Historical Overview," *International Labour Review*, September-October 1987; Charles Arden-Clarke, WWF-International, "Environment, Competitiveness and Countervailing Measures," presented to OECD Workshop on "Environmental Policies and Industrial Competitiveness," Paris, January 28-29, 1993.

72. Seaborne freight trade tonnage from Fearnresearch, op. cit., note 8; energy used in shipping is a Worldwatch Institute estimate, based on ibid. and on Stacy Davis and Melissa Morris, *Transportation Energy Data Book: Edition 12* (Oak Ridge, Tenn.: Oak Ridge National Laboratory, 1992); energy equivalents from U.S. DOE, op. cit., note 8; airfreight energy use is a Worldwatch Institute estimate based on data supplied by the International Civil Aviation Administration, Quebec, Canada, and by Larry C. Benton, Avmark, Inc., Arlington, Va., private communication, September 23, 1992; Marcia D. Lowe, "Rediscovering Rail," in Lester R. Brown et al., *State of the World 1993* (New York: W.W. Norton & Co.: 1993).

73. Task Force, op. cit., note 44; USTR, op. cit., note 44.

74. Task Force, op. cit., note 44; USTR, op. cit., note 44; H. Landis Gabel, INSEAD, Fontainbleau Cedex, France, "The Implications of International Trade Liberalization for the Environment: The Case of the Transport Sector," OECD, Paris, forthcoming; Gregory S. Johnson, "Shippers, Truckers Hope NAFTA Cuts Time, Losses," *Journal of Commerce*, August 12, 1992, Lowe, op. cit., note 72.

75. Task Force, op. cit., note 44; Gene M. Grossman and Alan B. Krueger, "Environmental Impacts of a North American Free Trade Agreement," Working Paper No. 3914, National Bureau of Economic Research, Cambridge, Mass., November 1991; U.S. Environmental Protection Agency (EPA), *1990 Toxics Release Inventory Public Data Release* (Washington, D.C.: 1992); Grossman and Krueger's estimate is most likely on the low side, as it assumes that the industries in Mexico produce the same amount of pollution per unit of output of a given product as U.S. industries do.

76. OECD, *Environmental Data Compendium 1991* (Paris: 1991).

77. Swedish pulp production and chlorine use from K. Karlsson, Swedish Pulp and Paper Association, Stockholm, private communication, November 2, 1992.

78. Grossman and Krueger, op. cit., note 75.

79. $12 billion estimate from study cited in "Poor Odds, High Stakes," *The Economist*, June 27, 1992; $90 billion estimate from Ian Goldin and Dominique van der Mensbrugghe, *Trade Liberalisation: What's At Stake?* Policy Brief No. 5 (Paris: OECD Development Centre, undated).

80. Peter Weber, "Green Seals of Approval Heading to Market," *World Watch*, July/August 1990; Ray V. Hartwell III and Lucas Bergkamp, "Eco-labelling in Europe: New Market-Related Environmental Risks?" *International Environment Reporter*, September 23, 1992.

81. OTA, op. cit., note 17. The EC has a somewhat different set of rules than GATT and NAFTA. In the EC, the goal is mutual recognition of member-country environmental standards, meaning that a product that is legal for sale in one country should in theory be accepted throughout the Community. In GATT and NAFTA, the only requirement is that national producers be subject to the same rules as importers and that the rule not be a disguised trade barrier. Though the EC rule is in theory more restrictive, environmental measures can be a legitimate reason for exceptions to the "mutual recognition" rules. See Task Force, op. cit., note 44.

82. For a description of the controversy over bovine growth hormone, see OTA, op. cit., note 17; Nancy Dunne and Bernard Simon, "Canada-U.S. Beer War Gets Green Tinge," *Financial Times*, July 31, 1992; Hilary F. French, "Battle Over a Bottle," *World Watch*, March-April 1993; Hufbauer and Schott, op. cit., note 60.

83. David Wirth, Assistant Professor of Law, Washington and Lee University School of Law, Testimony before the Committee on Science, Space, and Technology, U.S. House of Representatives, Washington, D.C., September 30, 1992; Steve Charnovitz, "Trade Negotiations and the Environment," *International Environment Reporter*, March 11, 1992.

84. Many environmentalists worried that removing the ability to require that bottle types be officially approved would mean that bottles would be merely collected for recycling, rather than for refilling. However, this fear has not been realized to any appreciable degree, per Jesper Hostrup, NOAH, Copenhagen, private communication, May 29, 1991. "Commission of the European Communities v. Kingdom of Denmark—Case 302/86," *Report on Cases Before the Court, Vol. 8* (Luxembourg: Office for Official Publications of the European Communities, 1988); "Landmark EEC Court Case on Returnable Bottles Gives Boost to Environment," *ENDS Report*, September 1988.

85. "Commission v. Kingdom of Denmark," op. cit., note 84; OTA, op. cit., note 17; Charnovitz, op. cit., note 83; Steve Charnovitz, "GATT and the Environment: Examining the Issues," *International Environmental Affairs*, Summer 1992; Wirth, op. cit., note 83.

86. GATT, "United States—Restrictions on Imports of Tuna: Report of the Panel," Geneva, September 3, 1991; David Phillips, Earth Island Institute, "Statement on the Implications of the GATT Panel Ruling on Dolphin Protection and the Environment," U.S. House of Representatives, Subcommittee on Health and Environment, Washington, D.C., September 27, 1991; Eric Christensen and Samantha Geffin, "GATT Sets Its Net on Environmental Regulation: The GATT Panel Ruling on Mexican Yellowfin Tuna Imports and the Need for Reform of the International Trading System," *Inter-American Law Review*, Winter 1991-1992.

87. GATT, "Report of the Panel," op. cit., note 86; OTA, op. cit., note 17. Some analysts have suggested that the judges in the tuna-dolphin case used an extremely narrow interpretation of GATT's exceptions clause, Article XX. Charnovitz, for instance, argues that the drafting history of the GATT suggests that the GATT founders meant to cover import bans based on process. See Steve Charnovitz, "Exploring the Environmental Exceptions in GATT Article XX," *Journal of World Trade*, October 1991.

88. GATT, op. cit., note 86.

89. Lee Jenkins, Humane Society, private communication, October 28, 1992; Sanford Gaines, Deputy Assistant U.S. Trade Representative for Environment, private communication, November 12, 1992; "GATT Council Agrees to EC Request for Dispute Panel on U.S. Tuna Ban," *Inside U.S. Trade*, July 17, 1992.

90. David Clark Scott, "Stung by US Tuna Ban, Mexico Protects Turtles," *Christian Science Monitor*, May 14, 1992; Charnovitz, op. cit., note 64; "U.S. Bans Importing Fish Caught With Drift Nets," *Washington Post*, September 20, 1992; Christensen and Geffin, op. cit., note 86.

91. Frances Williams, "Asean Condemns Timber Labelling," *Financial Times*, November 6, 1992; "Vienna, Malaysia to Set Up Joint Panel to Look at Eco- Labels on Tropical Woods," *International Environment Reporter*, December 2, 1992; "Parliament Rescinds Tropical Wood Tax, Maintains Product Eco-Label Requirement," *International Environment Reporter*, December 16, 1992.

92. "European Community to Ask for GATT Dispute Settlement Panel on U.S. Auto Taxes," *Inside U.S. Trade*, February 12, 1993.

93. Gaines, op. cit., note 89, private communication, February 16, 1993; Nigel Dudley, Earth Resources Research, *Importing Deforestation: Should Britain Ban the Import of Tropical Hardwoods?* (Godalming, Surrey, U.K.: WWF-UK, 1991); Arden-Clarke, op. cit., note 18; Dewi Sartika, "Log Export Ban Gets the Chop," *Panoscope*, September 1992.

94. Todd Campbell, "B.C. Fishermen Pay a High Price for Free Trade," *The New Pacific*, Winter/Spring 1990; OTA, op. cit., note 17.

95. OTA, op. cit., note 17; Jeffrey R. Vincent, "The Tropical Timber Trade and Sustainable Development," *Science*, June 19, 1992; Edward B. Barbier et al., London Environmental Economics Centre, "Environmental Effects of Trade in the Forestry Sector," OECD, Paris, forthcoming; Arden-Clarke, op. cit., note 32; On Indonesia, see Robert Repetto, *The Forest for the Trees? Government Policies and the Misuse of Forest Resources* (Washington, D.C.: World Resources Institute, May 1988).

96. GATT, op. cit., note 17; OTA, op. cit., note 17.

97. J.O. Cameron et al., "Relationship Between Environmental Agreements and Instruments Related to Trade and Environment," U.N. Conference on Environment and Development, Research Paper No. 35, Geneva, February 1992; Peter L. Lallas, Daniel C. Esty & David J. van Hoogstraten, "Environmental Protection and International Trade: Toward Mutually Supportive Rules and Policies," *The Harvard Environmental Law Review*, Vol. 16, No. 2, 1992.

98. Governments of Canada, Mexico, and the United States of America, op. cit., note 67.

99. OTA, op. cit., note 17; Steve Charnovitz, "NAFTA: An Analysis of Its Environmental Provisions," *Environmental Law Reporter*, February 1993; Charnovitz, op. cit., note 83.

100. French, op. cit., note 69; David Marsh, "The Patchwork that Has to Be Ironed Out," *Financial Times*, May 28, 1992; Hartwell and Bergkamp, op. cit., note 80.

101. Lori Wallach, Public Citizen, Washington, D.C., Memorandum to Environmental, Health and Consumer Advocates, December 26, 1991 (analysis of December 20, 1991 Dunkel Draft); James Cameron, *The Uruguay Round's Technical Barriers to Trade Agreement*, (Gland, Switzerland: WWF-International, January, 1993); Charnovitz, "NAFTA", op. cit., note 99.

102. For thoughts on environmental reform of GATT, see John H. Jackson, University of Michigan School of Law, unpublished memorandum of November 7, 1991, on "Changing GATT Rules"; Christensen and Geffin, op. cit., note 86; Peter Uimonen and John Whalley, "Trade and Environment: Setting the Rules," draft, Institute for International Economics, Washington, D.C., July 1992; and Charles Arden-Clarke, *International Trade, GATT, and the Environment* (Gland, Switzerland: WWF, May 1992).

103. Governments of Canada, Mexico, and the United States of America, op. cit., note 67; Charnovitz, "NAFTA," op. cit., note 99.

104. "Parliament Proposes Two-Year Moratorium on GATT Decisions Affecting the Environment," *International Environment Reporter*, January 27, 1993.

105. Cameron et al., op. cit., note 97; Governments of Canada, Mexico, and the United States of America, op. cit., note 67.

106. Thomas A. Wathen, *A Guide to Trade and the Environment* (New York: Environmental Grantmakers Association and the Consultative Group on Biological Diversity, July 1992).

107. Lori Wallach, "Trading Away the Environment," *Pesticides and You*, June 1992; Cameron, op. cit., note 101; James Cameron and Halina Ward, Foundation for International Law and Development, *The Multilateral Trade Organization: A Legal and Environmental Assessment* (Gland, Switzerland: WWF-International, May 1992); Letter from Stewart J. Hudson, National Wildlife Federation, Washington, D.C. to Carmen Suro-Bredie, Assistant U.S. Trade Representative for the Environment, November 10, 1992; "U.S. Proposals Would Change GATT Texts on Health, Safety Standards," *Inside U.S. Trade*, December 25, 1992; Ambassador Warren Lavorel, U.S. Coordinator for Multilateral Trade Negotiations, Office of the United States Trade Representative, briefing for U.S. non-governmental organizations (NGOs) and the business community, February 11, 1992.

108. "Trade and the Environment," Statement of Senator Max Baucus, Chairman, International Trade Subcommittee, Committee on Finance, U.S. Senate, Washington, D.C., September 17, 1992; "U.S. Senator Urges New GATT Round to Consider Inclusion of Environmental Code," *International Environment Reporter*, November 6, 1991; Paula L. Green, "International Trade Talks Likely to Be Greener," *Journal of Commerce*, August 19, 1992; Mark Magnier, "Power of Environmentalists Called Trade Issue of '90s," *Journal of Commerce*, July 20, 1992; "U.S. Sees Environmental Reform of GATT as Post Uruguay Round Issue," *Inside U.S. Trade*, November 20, 1992.

109. Group on Environmental Measures and International Trade, Interim Report by the Chairman, Ambassador H. Ukawa (Japan), December 3, 1992; "GATT Group Agrees to Step Up Work on Trade, Environmental Coordination," *International Environment Reporter*, December 16, 1992; "GATT Committee

Continues Discussions on Melding Trade, Environment Matters," *International Environment Reporter*, February 10, 1993.

110. Agenda 21, op. cit. 6; Charles Arden-Clarke, WWF-International, "Trade and Environment Policies After UNCED: Reconciling the Irreconcilable," Presentation to the Global Environment Research Center, Imperial College, University of London, September 16, 1992; "GATT Environment Work Delayed By Dispute Over Unilateral Action," *Inside U.S. Trade*, November 13, 1992.

111. It should be noted, however, that what the governments were really aiming to do was not to encourage widespread adoption of environmental taxes and other policies to internalize costs, but rather to discourage government subsidies for environmental protection that might distort trade. OTA, op. cit., note 17; Sanford E. Gaines, "The Polluter-Pays Principle: From Economic Equity to Environmental Ethos," *Texas International Law Journal*, Summer 1991; David A. Wirth, School of Law, Washington and Lee University, Remarks at Conference on Environmental Quality and Free Trade: Interdependent Goals or Irreconcilable Conflict, School of Law, Lexington, Va., September 25, 1992; "Rio Declaration on Environment and Development," U.N. Conference on Environment and Development, Rio de Janeiro, June 3-14, 1992.

112. OTA, op. cit., note 17.

113. OECD, op. cit., note 7; Carmen Suro-Bredie, Assistant U.S. Trade Representative for the Environment, briefing for NGOs and the business community, February 11, 1993.

114. Cameron and Ward, op. cit., note 107; Hilary F. French, "Hidden Success at Rio," *World Watch*, September/October 1992; Steve Charnovitz, "The Environment vs Trade Rules: Defogging the Debate," *Environmental Law,* forthcoming.

115. Other prominent regional pacts include discussions among the Southeast Asian nations, ASEAN, and the Latin America countries, Mercosur. In addition, the "Enterprise for the Americas" initiative announced by U.S. President George Bush in 1990 envisioned the negotiation of NAFTA-like accords with many of the South American nations. It is also expected that some of these countries will join in the NAFTA in future years.

116. "Expanding Trade and Creating American Jobs," Remarks by Governor Bill Clinton, North Carolina State University, Raleigh, North Carolina, October 4, 1992; Thomas L. Friedman, "Clinton Says U.S. Will Act Fast on Trade Pact If...," *New York Times*, January 9, 1993.

117. Hudson and Prudencio, op. cit., note 69; Sierra Club, op. cit., note 67; "Mexican Official Rejects Supranational Enforcement on Environment," *Inside U.S. Trade*, December 11, 1992; Rodger Schlickeisen, Defenders of Wildlife, Letter to President-elect Clinton, December 15, 1992.

118. IIE, op. cit., note 5; "NAFTA and the U.S./Mexico Border Environment: Options for Congressional Action," Texas Center for Policy Studies, September 1992; Jan Gilbreath, "Financing Environmental and Infrastructure Needs on the Texas-Mexico Border: Will the Mexican-U.S. Integrated Border Plan Help?," *Journal of Environment and Development,* Summer 1992; Congressman Richard A. Gephardt, Address on the Status of the North American Free Trade Agreement,

Institute for International Economics, Washington, D.C., July 27, 1992; Senator Max Baucus, Policy Address, National Association of Manufacturers Special Forum on Trade and the Environment, Washington, D.C., August 12, 1992; Nancy Dunne, "Mexico May Impose Environmental Tax," *Financial Times*, July 24, 1992; Albert Fishlow, Sherman Robinson, Raul Hinojosa et al., "Proposal for the Establishment of a North American Development Bank," University of California, Berkeley, January 1993; "New Fund to Provide 400 Million ECUs Through 1995 for Environmental Efforts," *International Environment Reporter*, August 12, 1992; USTR, op. cit., note 44; William K. Reilly, Administrator, EPA, Testimony before the Committee on Ways and Means, U.S. House of Representatives, Washington, D.C., September 15, 1992.

119. E.C. Delegation to the United States, *A Guide to the European Community* (Washington, D.C.: 1991).

120. French, op. cit., note 69; Keyes, op. cit., note 69; Sands, op. cit., note 69; "Monitoring the Implementation of Community Law on the Environment: An Initial Report by the Commission," Information Memo, Commission of the European Communities, February 8, 1990; "The Dirty Dozen," *The Economist*, July 20, 1991; David Wilkinson, "Greening the Treaty: Strengthening Environmental Policy in the Treaty of Rome," Institute for European Environmental Policy, London, October 1990; "New Fund to Provide 400 Million ECUs Through 1995 for Environmental Efforts," *International Environment Reporter*, August 12, 1992.

121. Cynthia Pollock Shea, "European Environmental Policy: Effects of the Single Market," *International Environment Reporter*, January 13, 1993; Verhoeve, Bennett, and Wilkinson, op. cit., note 4; Ralph Hallo, *How Green is the Treaty? The Environmental Provisions of the Treaty on European Union (Maastricht)* (Brussels: European Environmental Bureau, July 1992).

122. Basil Davidson, *The African Genius: An Introduction to African Social and Cultural History* (Boston: Atlantic Monthly Press, 1969); R.R. Palmer and Joel Colton, *A History of the Modern World* (New York: Alfred A. Knopf, 1978).

123. Robert Repetto et al., *Wasting Assets: Natural Resources in the National Income Accounts* (Washington, D.C.: World Resources Institute, 1989); Cruz and Repetto, op. cit., note 41; Alan B. Durning, *Poverty and the Environment: Reversing the Downward Spiral*, Worldwatch Paper 92 (Washington, D.C.: Worldwatch Institute, November 1989).

124. John Fitzgerald, Defenders of Wildlife, private communication, January 19, 1993; Richard B. Stewart, "International Trade and the Environment: Lessons from the Federal Experience," *Washington and Lee Law Review*, Fall 1992; Roosevelt quote cited in Brown, Goold, and Cavanagh, op. cit., note 68; "Southeast Asia Joins the Bloc Party: Six Nations Figure Best Trade Defense is a Zone of Their Own," *Washington Post*, November 10, 1992; Emily Thornton, "Will Japan Rule a New Trade Bloc?", *Fortune*, October 5, 1992.

125. Walter Russell Mead, "Bushism, Found: A second-term agenda hidden in trade agreements," *Harper's Magazine*, September 1992.

126. Kane, op. cit., note 1; Lester R. Brown, "Population Growth Sets Record," in Brown, Flavin, and Kane, op. cit., note 1.

THE WORLDWATCH PAPER SERIES

No. of
Copies

_____ 57. **Nuclear Power: The Market Test** by Christopher Flavin.
_____ 58. **Air Pollution, Acid Rain, and the Future of Forests** by Sandra Postel.
_____ 60. **Soil Erosion: Quiet Crisis in the World Economy** by Lester R. Brown and Edward C. Wolf.
_____ 61. **Electricity's Future: The Shift to Efficiency and Small-Scale Power** by Christopher Flavin.
_____ 62. **Water: Rethinking Management in an Age of Scarcity** by Sandra Postel.
_____ 63. **Energy Productivity: Key to Environmental Protection and Economic Progress** by William U. Chandler.
_____ 65. **Reversing Africa's Decline** by Lester R. Brown and Edward C. Wolf.
_____ 66. **World Oil: Coping With the Dangers of Success** by Christopher Flavin.
_____ 67. **Conserving Water: The Untapped Alternative** by Sandra Postel.
_____ 68. **Banishing Tobacco** by William U. Chandler.
_____ 69. **Decommissioning: Nuclear Power's Missing Link** by Cynthia Pollock.
_____ 70. **Electricity For A Developing World: New Directions** by Christopher Flavin.
_____ 71. **Altering the Earth's Chemistry: Assessing the Risks** by Sandra Postel.
_____ 73. **Beyond the Green Revolution: New Approaches for Third World Agriculture** by Edward C. Wolf.
_____ 74. **Our Demographically Divided World** by Lester R. Brown and Jodi L. Jacobson.
_____ 75. **Reassessing Nuclear Power: The Fallout From Chernobyl** by Christopher Flavin.
_____ 76. **Mining Urban Wastes: The Potential for Recycling** by Cynthia Pollock.
_____ 77. **The Future of Urbanization: Facing the Ecological and Economic Constraints** by Lester R. Brown and Jodi L. Jacobson.
_____ 78. **On the Brink of Extinction: Conserving The Diversity of Life** by Edward C. Wolf.
_____ 79. **Defusing the Toxics Threat: Controlling Pesticides and Industrial Waste** by Sandra Postel.
_____ 80. **Planning the Global Family** by Jodi L. Jacobson.
_____ 81. **Renewable Energy: Today's Contribution, Tomorrow's Promise** by Cynthia Pollock Shea.
_____ 82. **Building on Success: The Age of Energy Efficiency** by Christopher Flavin and Alan B. Durning.
_____ 83. **Reforesting the Earth** by Sandra Postel and Lori Heise.
_____ 84. **Rethinking the Role of the Automobile** by Michael Renner.
_____ 85. **The Changing World Food Prospect: The Nineties and Beyond** by Lester R. Brown.
_____ 86. **Environmental Refugees: A Yardstick of Habitability** by Jodi L. Jacobson.
_____ 87. **Protecting Life on Earth: Steps to Save the Ozone Layer** by Cynthia Pollock Shea.
_____ 88. **Action at the Grassroots: Fighting Poverty and Environmental Decline** by Alan B. Durning.
_____ 89. **National Security: The Economic and Environmental Dimensions** by Michael Renner.
_____ 90. **The Bicycle: Vehicle for a Small Planet** by Marcia D. Lowe.
_____ 91. **Slowing Global Warming: A Worldwide Strategy** by Christopher Flavin
_____ 92. **Poverty and the Environment: Reversing the Downward Spiral** by Alan B. Durning.
_____ 93. **Water for Agriculture: Facing the Limits** by Sandra Postel.
_____ 94. **Clearing the Air: A Global Agenda** by Hilary F. French.
_____ 95. **Apartheid's Environmental Toll** by Alan B. Durning.
_____ 96. **Swords Into Plowshares: Converting to a Peace Economy** by Michael Renner.
_____ 97. **The Global Politics of Abortion** by Jodi L. Jacobson.
_____ 98. **Alternatives to the Automobile: Transport for Livable Cities** by Marcia D. Lowe.
_____ 99. **Green Revolutions: Environmental Reconstruction in Eastern Europe and the Soviet Union** by Hilary F. French.

_____100. **Beyond the Petroleum Age: Designing a Solar Economy** by Christopher Flavin and Nicholas Lenssen.
_____101. **Discarding the Throwaway Society** by John E. Young.
_____102. **Women's Reproductive Health: The Silent Emergency** by Jodi L. Jacobson.
_____103. **Taking Stock: Animal Farming and the Environment** by Alan B. Durning and Holly B. Brough.
_____104. **Jobs in a Sustainable Economy** by Michael Renner.
_____105. **Shaping Cities: The Environmental and Human Dimensions** by Marcia D. Lowe.
_____106. **Nuclear Waste: The Problem That Won't Go Away** by Nicholas Lenssen.
_____107. **After the Earth Summit: The Future of Environmental Governance** by Hilary F. French.
_____108. **Life Support: Conserving Biological Diversity** by John C. Ryan.
_____109. **Mining the Earth** by John E. Young.
_____110. **Gender Bias: Roadblock to Sustainable Development** by Jodi L. Jacobson.
_____111. **Empowering Development: The New Energy Equation** by Nicholas Lenssen.
_____112. **Guardians of the Land: Indigenous Peoples and the Health of the Earth** by Alan Thein Durning.
_____113. **Costly Tradeoffs: Reconciling Trade and the Environment** by Hilary F. French.

_____ **Total Copies**

☐ **Single Copy: $5.00**
☐ **Bulk Copies (any combination of titles)**
☐ 2–5: $4.00 ea. ☐ 6–20: $3.00 ea. ☐ 21 or more: $2.00 ea.

☐ **Membership in the Worldwatch Library: $25.00 (international airmail $40.00)**
The paperback edition of our 250-page "annual physical of the planet," *State of the World 1993,* plus all Worldwatch Papers released during the calendar year.

☐ **Subscription to *World Watch* Magazine: $15.00 (international airmail $30.00)**
Stay abreast of global environmental trends and issues with our award-winning, eminently readable bimonthly magazine.

No postage required on prepaid orders. Minimum $3 postage and handling charge on unpaid orders.

Make check payable to Worldwatch Institute
1776 Massachusetts Avenue, N.W., Washington, D.C. 20036-1904 USA

Enclosed is my check for U.S. $_______

____

name **daytime phone #**

____

address

____

city **state** **zip/country**